## DATE DUE

| | |
|---|---|
| | |
| | |
| | |
| | |
| | |
| | |
| | |
| | |
| | |
| | |
| | |
| | |
| | |
| | |
| | |
| | |
| | |
| | |

# THE BIOENGINEERED FOREST

*Challenges for Science and Society*

# THE BIOENGINEERED FOREST

## Challenges for Science and Society

EDITED BY
STEVEN H. STRAUSS
AND H. D. (TOBY) BRADSHAW

RESOURCES FOR THE FUTURE
WASHINGTON, DC, USA

An RFF Press book
Published by Resources for the Future
1616 P Street, NW
Washington, DC 20036–1400
USA
www.rffpress.org

**Library of Congress Cataloging-in-Publication Data**

The bioengineered forest : challenges for science and society / Steven H. Strauss and H.D. Bradshaw, editors.
    p.   cm.
    Includes bibliographical references.
    ISBN 1–891853–71–6 (hardcover : alk. paper)
    1. Forestry biotechnology. 2. Trees—Biotechnology. 3. Forest products—Biotechnology. I. Strauss, Steven H., 1955–   II. Bradshaw, H. David.
SD387.B55B56    2003
634.9—dc22                                              2003018033


f e d c b a

The paper in this book meets the guidelines for permanence and durability of the Committee on Production Guidelines for Book Longevity of the Council on Library Resources.

This book was designed and typeset in Minion by Betsy Kulamer. It was copyedited by Nancy Geltman. The cover was designed by Maggie Powell. Cover art: *La belle saison,* 1950, by René Magritte, ©2004 C. Herscovici, Brussels/Artists Rights Society (ARS), New York. Photo credit: Banque d'Images, ADAGP/Art Resource, New York.

ISBN 1–891853–71–6

# *About* Resources for the Future *and* RFF Press

**R**esources for the Future (RFF) improves environmental and natural resource policymaking worldwide through independent social science research of the highest caliber. Founded in 1952, RFF pioneered the application of economics as a tool for developing more effective policy about the use and conservation of natural resources. Its scholars continue to employ social science methods to analyze critical issues concerning pollution control, energy policy, land and water use, hazardous waste, climate change, biodiversity, and the environmental challenges of developing countries.

**RFF Press** supports the mission of RFF by publishing book-length works that present a broad range of approaches to the study of natural resources and the environment. Its authors and editors include RFF staff, researchers from the larger academic and policy communities, and journalists. Audiences for RFF publications include all of the participants in the policymaking process—scholars, the media, advocacy groups, nongovernment organizations, professionals in business and government, and the general public.

# Contents

## PART I
### *Economic and Technological Choices*

# PART II
## *Ethical, Social, and Ecological Caveats*

# Preface

The *Bioengineered Forest* had its origins in 1999, as activist outcry over agricultural biotechnology was growing loud around the globe. We, two researchers in the United States, were feeling confident of the strength of our field's biotechnology and how it would stand in a public forum. We embarked on organizing a symposium in which representatives of all positions on forest biotechnology could be heard and in which a path forward, for research and for social deliberations, could be charted.

We chose to limit the scope of the symposium to bioengineered plantations—those employing trees whose genetic makeup had been directly altered via recombinant DNA methods. These kinds of trees are also known, in various parts of the world, as genetically engineered, genetically modified, or transgenic. Genetic modification comprises a small subset of the diverse tools that come under the rubric of "forest biotechnology," but it is the most controversial, and, in terms of the potential for novel changes to trees that could not be accomplished via traditional breeding, it is the most powerful.

To provide a breadth of views, we included presenters from environmental organizations and scholars with critical views of bioengineered trees. The atmosphere of the meeting was charged. Vandalism and arson attacks against our own research had occurred just weeks before, and serious damage was done to laboratories and field plots. News articles about so-called ecoterrorism in the Pacific Northwest appeared in *Science* magazine in April (292, 34–35) and June (292, 1622–1633) 2001 and in most newspapers. A peaceful demonstration was staged the afternoon that the meeting began. Numerous county, state, and federal security agents were present to ensure that the planned open exchange of ideas could proceed without disruption.

Held during July 2001 at Skamania Lodge, in Stevenson, Washington, the meeting attracted more than 200 attendees from 23 countries and was widely regarded as a great success. Nearly all of the 28 speakers, who included many of the most active scientists and thoughtful scholars in the field, delivered high-quality papers. Half had no significant ties to industrial or corporate interests and presented broad ethical, ecological, or environmental views. The majority of funding for the meeting came from a competitive grant from the U.S. Department of Agriculture (USDA) Biotechnology Risk Assessment Program, though a number of other public agencies and companies also contributed small amounts.

* * * * *

This book contains a revised and updated selection of papers presented at the symposium, plus additional contributions chosen to give a balance and diversity of views. We regret that space limits prevented inclusion of the papers of David Victor and Jessie Ausubel, W. Stephen Burke, Sandy Thomas, David S. Heron and John L. Keough, Sue Mayer, James R. Boyle, Helen Lundkvist and C. T. Smith, and Faith T. Campbell and Rachel Asante-Owusu.

The first part of the book deals with the broad context of forestry, the technological possibilities, and the research needed to allow responsible use of genetically engineered plantation trees. The second part comprises discussions of the pitfalls, costs, and diverse risks of impetuous, poorly planned, or socially undesirable applications. Interestingly, none of the authors who presented at the conference felt that there should be a moratorium on research or that genetically engineered trees should be excluded from all forest plantations. The discussions instead revolved around the kinds of new varieties of trees, how well studied, how perceived by the public, and with what legal and social license.

A clear message from the first half of the book is that a tremendous amount of research is needed to develop the technology. The private sector, however, is unlikely to be able to move the field forward on its own, both because most forestry companies have only limited interest in funding research in this area and because research information developed by corporations has limited credibility, especially in the environmental sector. A comparably clear message from the second half of the book is that the experience to date in agricultural biotechnology tells us far more about what not to do than about what to do in commercializing forest biotechnology. It calls for up-front design of gene constructs, with environment and sustainability in mind; thorough, multidisciplinary testing, through to field deployment, of economic benefits and environmental effects; and much more work to obtain social license for the products. Neither "save the forests" shibboleths nor hiding environmentally ill-advised products behind high-profile products such as trees resistant to exotic diseases will serve society's needs. Opinion polls show that the public has the ability to discriminate good from bad uses of technology, that it takes both the doom cries of advocacy groups and the hype of corporations with a grain of salt, and that it forms opinions with discrimination. Though most Americans approve of crop biotechnology, the majority is not a large one; the minority that are negative have strong reservations, and approval ratings vary widely among applications. Nonetheless, the public is willing to accept risks if the benefits and uncertainties to society are presented in clear, sober language and are backed by credible research. Public education about the science behind genetically engineered organisms is therefore critical so that informed decisions can be made in the political arena.

A careful reading of the chapters in this book reveals several key themes.

**Communication challenges are substantial.** Science and business have clearly lost the first round in the communication wars over crop biotechnology. If people are to accept well-considered applications of tree biotechnology, we need to create new approaches to establishing trust. Thompson suggests that the plant biotechnology profession has done a poor job of communicating what it does, and why, to the public and of explaining the professional ethic that guides it. He states that "by absenting themselves from any discussion of the social networks in which their work is applied and the technologies that are adapted from it, life scientists have adopted an ethic that permits powerful actors to use science silently in the extension and exertion of their power." He goes on, "If uncertainties and ethical silences are allowed to linger, the public will be entirely justified in regarding tree biotechnology with the same skepticism and hostility that now plague agricultural and food applications." Thompson calls for much more vigorous articulation of the rationale for tree biotechnology by its practitioners.

Intense rhetoric and polarization of views present great challenges to any communication about biotechnology. As Hal Salwasser points out in Chapter 1, "Extremist ideologies abound. They feed on controversy, and where none exists, they create it. Such ideologies taken to the extreme destroy any potential for common ground and mutual progress."

**Trust and social control are key issues.** The close corporate control over the tools and products of bioengineering and their roles in furthering consolidation in life sciences companies have been key sources of resistance to agricultural biotechnology. As Thompson points out, "Many of the activists who have opposed biotechnology in agriculture ground their opposition in a sociopolitical argument." And he says, "[B]iotechnology became a focus for those who see technology as a force driving modern societies toward economic and political inequality."

The close alliances between academic science and industry, as well as the commercial interests of universities in benefiting from patents since changes in the laws governing university research, have also undermined trust in academic science and opinion. Furthermore, the forest industry does not enjoy great public trust in environmental matters. As Doering points out in Chapter 8, "The logging and timber products industry has . . . reputational liabilities for its poor past and current environmental record and political stands and activities. ... A union of the timber industry and the plant biotechnology industry comes poorly armed to a battle based on public trust and may be a dream marriage to an anti-corporate activist." The virtual rejection of crop biotechnology in Europe, which many attribute to a lack of trust in information provided by corporations, government, and academic scientists, underlines the importance of trust to public acceptance of new technologies, especially those whose benefits to consumers are indirect.

Although improving productivity via intensified plantation technology can theoretically allow society to grow more of its wood on less land, thus preserving more land for biodiversity and other ecological values, many social and ecological obstacles stand in the way. As Friedman and Charnley point out in Chapter 9, "If wood production becomes unprofitable for private landowners because the increased efficiency [of intensively managed plantations] depresses prices, they may sell the forestland they own or convert it to other uses that are more profitable, such as agriculture or residential development. For the public to ensure that current forestlands are maintained in forests, it may need to acquire land, create new tax incentives, or impose regulations." They also argue that establishment of plantation estates can bring social and ecological upheavals that are often poorly accounted for by industrial and government desires for plantation-based wood industries. In India, they note, "the planting of eucalyptus and other fast-growing exotics on these common lands has deprived local communities of that source of fodder and fuel, undermining their subsistence. This alienation of village common lands has generated great conflict, as industry now competes with local communities for land." Friedman and Charnley call for more localized and socially deliberated choices over forestry practices.

**Legal obstacles are great.** The costs and uncertainties of obtaining necessary government and private licenses are major reasons for the reluctance of many companies and public sector researchers to invest in bioengineering. Because bioengineered trees typically result from the use of multiple patented technologies, it is often necessary to obtain several patent licenses, a process complicated by the long time span for research and development and the difficulty of estimating patents' value given the many technical and social acceptance uncertainties. It is also not clear that obtaining a large patent estate, as many agricultural biotechnology companies have chosen to do, is a good strategy for tree biotechnology. Doering suggests, "The case for patenting trees is not obvious and merits analysis. Patenting may be a weak and incorrect strategic path for the forest biotechnology industry."

The United States has much less restrictive regulations governing field tests than Europe does, but obtaining approval to commercialize bioengineered trees is fraught with legal uncertainties. As Bryson and her colleagues point out, "Products of biotechnology do not always fit comfortably within the lines the law has drawn based on the historical function and intended use of products." They describe several case studies that "highlight the potential complexity of federal reviews of such products and the overlapping applicability of a number of the primary environmental and natural resource laws. The bioremediation poplar case study discusses the application and coordination of the Plant Protection Act, the Toxic Substances Control Act, and the Comprehensive Environmental Response, Compensation and Liability Act." They suggest that any decision by USDA's Animal and Plant Health Inspection Service on commercial release might require the filing of an environmental impact statement, under the provisions of the National Environmental Policy Act. Unless Congress simplifies the overlapping laws, even products expected to have large environmental benefits, such as trees that detoxify polluted soils, may be extremely difficult to commercialize.

**Benefits are large and diverse.** The authors of the following chapters cite a number of potential benefits of bioengineered trees. They include the basic agronomic goal of intensifying production in a limited area to reduce the amount of forestland required to supply society's wood needs. Salwasser believes, "We appear to be within a decade or two of a point at which approximately 40% of the industrial wood fiber produced in the world comes from planted forests. . . . If predictions by some of the leading scientists are correct, eventually only 10 to 15%—perhaps even less—of the world's forested area will be in planted forests managed for high yields of wood. . . ."

In Chapter 2, Lucier, Hinchee, and McCullough observe, "Besides improving the quality and quantity of raw material supplies, over the long term biotechnology could have a radical impact on pulping processes, waste-to-energy systems, and other aspects of forest products manufacturing." Sedjo points out in Chapter 3 that "the economic benefit associated with the introduction of only one transgene, the herbicide resistance gene, could be as much as $1 billion annually."

"The introduction of biotechnology to forestry has the potential to produce great economic benefits in the form of lower costs and increased availability to consumers of wood and wood products," says Sedjo. Others point out that increased efficiency in producing wood should promote its use as a renewable source of energy and materials. Lucier observes, "The forest products industry is already the world leader in biomass energy production and will probably increase its production substantially if new technologies such as biomass gasification are successful." Other benefits include new tools for overcoming invasive organisms and exotic pests, improved ability to breed trees for difficult environments, and better customization of wood for different products. Raffa suggests in Chapter 13 that "transgenic capabilities may . . . help us respond more quickly to invasive species, which are an increasing threat associated with heightened global commerce." Doering states, "Among the traits and benefits that forest biotechnology is interested in, the reduced-lignin designs may have the greatest social utility; they would reduce chemical, water, and energy use by the pulp and paper industry as well as the pollution it creates." Johnson and Kirby summarize it nicely in Chapter 12, saying, "Improvements to forest trees via conventional selective breeding are severely limited, but transgenic technology offers the opportunity to domesticate trees, so as to tailor their characteristics more closely to the requirements of commercial forestry and the end user of forest products."

**The risks are complex.** The risks of undesired effects from bioengineering derive from two sources: the addition of genes from distant species that impart novel functions, and alterations to native or homologous genes that cause large changes in gene expression. Both produce novel properties in trees, which can have significant social or ecological consequences. As Raffa says, "The issue of whether transgenic plants could exert serious environmental effects hinges greatly on whether there is gene flow into native plants." However, many kinds of plantations also support significant fauna that depend on pollen or fruits, and the removal of reproductive organs as a means to impede gene flow is thus problematic. As Johnson and Kirby point out, "The most obvious solution to maintaining biodiversity among sterile trees would be to plant mixed stands of transgenic and conventional trees, either in direct admixture or by planting blocks of different trees."

The long life cycle of trees complicates risk estimation but also provides buffers compared to annual crops. As Burdon and Walter describe in Chapter 5, the relatively long lifespan of trees can accentuate risks of any tree-improvement technology, in several ways. "A crop can be exposed to risk over a longer period. A delayed failure, which may still occur before profitable salvage is possible, will carry the compounded costs of crop establishment, tending, and protection. . . . On the other hand, if problems are observed during initial years of establishment, further plantings can be stopped, whereas with annual crops vast plantings may be converted to transgenic types in a few years, as occurred in the United States with glyphosate-tolerant soybeans."

The long lifespan of tree crops also makes it impossible to rely solely on empirical studies to evaluate risks and benefits; some kind of modeling is necessary. Johnson and Kirby believe, "With long-lived perennials such as trees . . . a predictive modeling approach is the only realistic option if regulatory authorities are to make decisions about release within a reasonable time." However, they also point out that such models are useless unless they contain reliable data about performance and fitness of bioengineered trees. "Without evidence from field testing and pilot-scale studies it is difficult to see how risk can be assessed in a way that is both scientifically defensible and publicly acceptable."

Bioengineered trees must perform stably, showing a minimum of unexpected physiological defects, and gene transfer and regeneration of transgenic plants must be feasible in diverse species and genotypes if they are to be put into broad commercial use. In Chapter 4, Meilan and his coauthors summarize a number of studies, the majority based on poplars, that showed predominantly stable transgene expression and low, manageable levels of unintended (somaclonal) variation: "It has been shown that we can produce transgenic trees with almost no evidence of collateral genetic damage and that inserted genes are expressed stably from year to year, after vegetative propagation, and in a variety of environments." They believe, however, that for most tree species and genotypes, the ability to produce a large number of transgenic products (events), to ensure that highly vigorous ones can be selected and that genetic diversity is not unduly restricted, is problematic and that "the main challenge with trees continues to be the development of efficient transformation systems for the most desired genotypes."

Transgenic expression, as Raffa says, offers "some unique opportunities for pest resistance in plantation trees that cannot be achieved easily by other means." However, they also raise problems for pest management and prediction of nontarget effects. "A major challenge lies in the inability of ecologists to quantify the likelihoods of various potential adverse effects, given the requisite long time frames, complex trophic interactions, and extensive spatial scales."

Hancock and Hokanson (Chapter 11) consider the potential for genetic engineering to make species more invasive. Considering the record of intensive breeding and hybridization as an indication of the effects of genetic engineering, they observe, "Even though crop species have been planted among their progenitors for thousands of years, we are not aware of any report where the native fitness of the wild species was noticeably changed." They argue that the analogy between invasive organisms and transgenic crops is largely a false one: "[A] high percentage of the exotic species that become invasive are already excellent colonizers somewhere else, and their population size explodes when they are introduced into a new area where there are few to none of the natural constraints with which they evolved. This is very different from the situation facing transgenic forestry and agronomic crops." Hancock and Hokanson argue persuasively that "it is much easier to predict the environmental risk of transgenic trees than an exotic introduction, as the level of risk in transgenics can be measured by evaluating the fitness impact of a single engineered trait, rather than a whole syndrome of potentially invasive traits." Nevertheless, Raffa amply shows that assessing the ecological risk of a transgene that may increase tree fitness and have significant nontarget effects in wild relatives, which may take decades to centuries to manifest themselves, is by no means easy.

Modifications of native gene expression also raise complex ecological questions. For example, as Johnson and Kirby point out, "Transgenic technology offers a way of achieving radical changes in the lignin-to-cellulose ratio of both conifers and deciduous hardwoods. Doing that, however, has the potential inadvertently to alter the susceptibility of such trees to animal and fungal attack, with consequent damage to the trees." Many scientists believe that such modifications, which typically employ altered versions of native or homologous genes, do not require limits to gene flow because the progeny should be at a competitive disadvantage where wild trees are abundant. Strauss and Brunner state that "'domestication' alleles that may have large benefits for productivity or product quality under intensive plantation management . . . are virtually inaccessible to traditional breeding because of their adverse effects on fitness in wild populations." Doering, however, disagrees, saying, "Alteration of fundamental properties such as lignin content and growth rates will also demand mechanisms to guarantee no flow of potentially deleterious traits to wild tree ecosystems."

\* \* \* \* \*

After consideration of all the views set out in this volume, and as participants and observers in the debates over crop biotechnology of the last few years, we remain optimistic and intellectually motivated about the possibilities for bioengineering of plantation trees to improve productivity, to help trees threatened by pests, and to produce—with rational deployment—large net social and environmental goods. It is clear, however, that many people and organizations disagree with the agronomic paradigm for forestry or believe that the products that bioengineering can deliver at this early stage of its development are not worth the risks they entail. Many are also fundamentally suspicious of the dominant corporate presence and large role for intellectual property rights associated with bioengineering. Large numbers of people appear unwilling to tolerate the infusion of novel kinds of organisms into the environment when the agenda is controlled by large agrichemical and forestry companies whose ethics in regard to environmental stewardship are suspect and whose operations are not transparent—public relations efforts notwithstanding. Many who hold views such as these make a sound case for ethical hesitation and show admirable respect for biodiversity and the complexity of nature. They urge a greater role for social choice concerning technologies, instead of simple scientific reductionism and industrial efficiency.

Despite our own enthusiasm for bioengineering science, this book and the symposium that preceded it have shown us that for the foreseeable future, most commercial uses of bioengineered plantations face difficult odds, at least in the developed world. The scientific potentials are vast, but society and business have posed large obstacles to rapid progress toward application. It is our hope that the thoughtful discussions in these chapters will encourage leaders in government, business, and public organizations to work toward reducing polarization in public debate, toward constructing scientifically rational regulations that do not penalize useful innovations, and toward more partnerships in research and development between the public and private sectors. These steps would help to bring the rapid scientific advances of this age of genomics out of the laboratory and into the light. We are convinced that our continually growing, resource-limited, and changing world will need tree biotechnology. We worry whether society will make the needed investments and adjustments in time for it to matter.

STEVEN H. STRAUSS
H. D. (TOBY) BRADSHAW

# Acknowledgments

We wish to thank the many people and organizations that made this book possible. Jace Carson, the do-it-all-and-smile program manager who worked at Oregon State University from 1998 to 2003, managed most aspects of the symposium from which the book derives, as well as preparation of the manuscripts for the book. The U.S. Department of Agriculture Biotechnology Risk Assessment Grants Program provided major support for the symposium. A number of companies and public agencies also provided support for speakers and student scholarships for the symposium. Colleagues too numerous to name, from academia, public agencies, and industries, have supported and stimulated us from the impressive growth phases of our tree biotechnology programs, through the dark days of "ecoterrorism," and to the era of controversy and company retreat from research in which we now find ourselves. It has been a privilege to collaborate with them all.

We thank Tom Ledig, the conservation geneticist who nudged the traditional world of forest genetics toward molecular genetics and biotechnology in the early 1980s. We thank Bill Libby, for never being satisfied with what is routine in forest genetics. We thank Reini Stettler, whose infectious enthusiasm for the biology and genetics of forest trees inspired several generations of scientists to take up the challenge of studying our planet's largest and oldest living things. We thank David Neale, who has fostered the development of a real sense of community among forest geneticists, and without whose leadership the entire field would be chaotic, at best.

We thank Rick Meilan, Amy Brunner, Caiping Ma, Jace Carson, and Steve DiFazio for their daily above-and-beyond-the-call efforts in sustaining bioengineering research in the Pacific Northwest. We thank the numerous industrial and public agency colleagues with whom we have productively worked over the years, particularly Jake Eaton, Ray Ethel, Larry Miller, Rob Miller, Peggy Payne, Brian Stanton, Jerry Tuskan, John Trobaugh, Nick Wheeler, and Cees Van Oosten.

Finally, we thank all the speakers and participants at the extraordinary symposium that gave rise to this book. Biology is the most complex science on Planet Earth. Biotechnology, with all its possibilities, novelties, and risks, seems to be the most complex technology. We set aside two days for deliberation. All indications are that the deliberations will continue on so long as there is human ingenuity, need, and a natural world to sustain.

We dedicate this book to our mentors and colleagues in forest genetics and biotechnology, of whom we have had space to name only a few. Never have we experienced a community of scholars that is more engaging, mutually supporting, and sincerely interested in human progress and environmental conservation.

# Contributors

**H. D. (TOBY) BRADSHAW** is a professor in the Department of Biology at the University of Washington. His research interests are in the genetic basis of adaptive trait evolution in natural populations. With a background in molecular biology, he has collaborated with ecologists, physiologists, and evolutionary biologists on many research projects and produced publications on genetic adaptation.

**AMY M. BRUNNER** is an assistant professor in the Department of Forest Science at Oregon State University. Her research focuses on the molecular genetics of tree development, use of genetic engineering to modify tree maturation and flowering, and applications of genomics to forest biology. Her recent publications include "Structure and Expression of Duplicate AGA-MOUS Orthologs in Poplar" and "Controlling Maturation and Flowering for Forest Tree Domestication."

**NANCY S. BRYSON** is the general counsel of the U.S. Department of Agriculture. Prior to her nomination and confirmation to this position, she was a partner with Crowell & Moring's Natural Resources and Environment division and co-chair of the firm's biotechnologies practice.

**ROWLAND D. BURDON** has been a scientist at the New Zealand Forest Research Institute since 1964. His main research has addressed the genetic resources of Monterey pine (*Pinus radiata*), quantitative breeding methodology, and tree breeding strategy, and in recent years he has focused on risk management for genetic improvement.

**SUSAN CHARNLEY** is a research social scientist with the USDA Forest Service, Pacific Northwest Research Station, in Portland, Oregon. She has been the national program leader for Human Dimensions in the National Forest Systems, based in Washington, D.C. Her research has focused on the social dimensions of resource use in communities in the United States, Central America, and East Africa.

**DON S. DOERING** is a senior associate at Winrock International and has biotechnology experience through work in venture capital, strategic and technical consulting, entrepreneurship, and strategy studies. His expertise is in corporate, science, and public policy relating to genetic engineering and sustainable agriculture. He was an associate at Calvert Social Venture Partners, founding vice president of AquaPharm Technologies Corporation, a senior fellow at the Wharton School of Business, and a senior associate at World Resources Institute. He serves on the Biotechnology Advisory Council of Monsanto, the Institute for Forest Biotechnology, and the North American Commission for Environmental Cooperation Maize Advisory Group.

**DAVID ELLIS** is a plant physiologist at the Seed Viability and Storage Research Unit of the National Center for Genetic Resources Preservation, USDA Agricultural Research Service. He heads a group researching cryo-preservation of vegetatively propagated crops to ensure the long-term survival and availability of genetic material for future generations. His current focus is on grapes, sweet potato, and garlic.

**SHARON T. FRIEDMAN** is the national research program leader for genetics and silviculture with the USDA Forest Service. She has published on gene flow in forest trees, forest tree breeding, and the application of science to policy. In 2000, she was the science lead of a White House assessment of the efficacy of domestic regulation of genetically engineered organisms released into the environment. She is currently chair of the Forest Science and Technology Board at the Society of American Foresters, and is a fellow of the Society.

**JAMES F. HANCOCK** is a professor of horticulture at Michigan State University. His interests include evolutionary genetics, breeding of small fruits, and the environmental ramifications of transgene escape. He has participated in numerous conferences on the environmental impacts of genetically engineered plants. His books include *Plant Evolution and the Origin of Crop Species.*

**MAUD HINCHEE** is the chief technology officer for ArborGen, LLC, a company that researches commercialization of genetically improved trees for sustainably managed plantation forests. Before joining ArborGen, she had 18 years experience in developing technology to produce genetically improved agricultural crops. She is currently a board member of the Institute for Forest Biotechnology and an editor for the *Plant Biotechnology Journal.*

**KAREN HOKANSON** is a research associate in the department of horticultural science at the University of Minnesota. Her interests include molecular plant ecology and risk assessment of transgenic crops. She has served as staff scientist/biotechnologist with the USDA/APHIS/PPQ and has participated in over a dozen international meetings on the regulation of GMOs and biotechnology.

**BRIAN JOHNSON** is senior adviser on biotechnology to the British statutory nature conservation agencies and is head of the Agricultural Technologies Group at English Nature, one of the U.K. government's advisers on nature conservation. He has written numerous articles in the scientific and popular press about conservation and the impact of biotechnology on the environment and sits on several U.K. advisory committees concerned with genetics research, regulating the release of GMOs into the environment, and the development of sustainable farming methods.

**KEITH KIRBY** has been a forestry and woodland officer for 24 years with English Nature, a government agency concerned with wildlife protection in England. He has published many papers and a book about the status and conservation of native woodland in the U.K. and is widely recognized as a leading authority on natural forest and plantation ecology.

**ALAN A. LUCIER** is senior vice president with the National Council for Air and Stream Improvement, and a cofounder of the Institute of Forest Biotechnology. His primary research interests are environmental and technological aspects of sustainable forestry. He is a member of the National Commission on Science for Sustainable Forestry and editor of the book *Mechanisms of Forest Response to Acidic Deposition.*

**RICHARD J. MANNIX** is counsel at Crowell & Moring LLP. His legal practice is devoted to assisting clients with environmental, health, and safety issues, with particular emphasis on pesticides, toxic substances, and clean air regulation. Before joining Crowell & Moring in 1988, he served as director of legislative and regulatory affairs for a major public utility and was senior editor of the *Brooklyn Law Review.*

**REX B. MCCULLOUGH**, former vice president of forestry research at Weyerhaeuser, recently retired after 25 years with the company.

**RICK MEILAN** is an associate professor at Purdue University and was formerly associate director of the Tree Genomics and Biosafety Research Cooperative based at Oregon State University. His recent research has focused on genetic engineering of poplar and precocious flowering of woody angiosperms. Recent publications include "Modification of Flowering in Transgenic Trees" and "Stability of Herbicide Resistance and GUS Expression in Transgenic Hybrid Poplars."

GILLES PILATE is research scientist at the Unité Amélioration, Génétique et Physiologie Forestières (INRA, Orléans, France). His research interest is in molecular physiological mechanisms of wood formation. He has authored a number of scientific papers, including: "Transformation in Larch (*Larix species*)"; "Transgenic Poplar Trees (*Populus species*)"; and "Genetic Engineering of Poplar Lignins."

STEVEN P. QUARLES is chair of the natural resources and environmental law group of Crowell & Moring LLP. He represents and litigates on behalf of forest products, mining, agriculture, and residential and commercial development trade associations and companies on environmental matters. He has served as deputy undersecretary in the Department of the Interior and counsel to the Senate Committee on Energy and Natural Resources.

KENNETH F. RAFFA is professor of forest entomology at the University of Wisconsin–Madison. He researches population dynamics of forest insects, particularly the roles of tree defenses and natural enemies. An associate editor of *Ecology* and a former associate editor of *Forest Science*, he has also served on the National Research Council Panel on the Future of Pesticides and many other grant panels.

HAL SALWASSER is professor of forest resources and forest science, dean of the College of Forestry, director of Oregon's Forest Research Laboratory, and director of the Institute for Natural Resources at Oregon State University. Prior to his work at Oregon State, he has served in a variety of positions at the USDA Forest Service. He is a past president of the Wildlife Society and the author of over 70 research papers and book chapters.

ROGER A. SEDJO is senior fellow and director of the forest economics and policy program at Resources for the Future. His recent research has focused on the economics of biotechnology in forestry. Earlier work included research on the economics of timber supply and plantation forests, biodiversity and bioprospecting, and climate change and forestry. His most recent book is *A Vision for the U.S. Forest Service*.

JEFF SKINNER is a postdoctoral faculty research associate in the department of horticulture at Oregon State University. His research interests include genes controlling floral development, methods to manipulate tree flowering, and genes contributing to abiotic stress tolerance. He has published several papers on floral genes and methodologies for engineering sterility in poplar trees, including a review paper entitled "Options for Genetic Engineering of Floral Sterility in Forest Trees."

**STEVEN H. STRAUSS** is a professor in the department of forest science at Oregon State University where he is director of the Tree Genomics and Biosafety Research Cooperative. He has published more than 100 scientific papers and given more than 130 invited lectures on genetics, evolution, and biotechnology of trees. He served on a number of panels at the U.S. National Research Council, National Science Foundation, and Department of Agriculture. His current research focuses on genomic analysis and genetic engineering methods for trees, and he directs a university–industry consortium that is developing genetic engineering solutions to mitigate concerns of gene flow from transgenic plantations.

**PAUL B. THOMPSON** is the author of *Food Biotechnology in Perspective* and is internationally known for his work on the environmental ethics of genetic engineering. He holds the W.K. Kellogg Chair in Agricultural, Food, and Community Ethics at Michigan State University.

**CHRISTIAN WALTER** is a senior scientist at the New Zealand Forest Research Institute. His research over the last decade has focused on the development and application of genetic engineering technology to conifer tree species. He is currently editing a book titled *Modern Plantation Forest Biotechnologies for the 21st Century.*

# PART I

# *Economic and Technological Choices*

# 1

# Future Forests

## *Environmental and Social Contexts for Forest Biotechnologies*

### Hal Salwasser

W hy are people concerned about forests and biotechnology? As we contemplate the social and ethical implications of biotechnology in general, and genetically modified forest organisms in particular, we start from the premise that forests are vitally important to human well-being. Given the growing human population, shrinking global forest area, and pressures on forests to meet many kinds of human needs and provide ecosystem services, biotechnology may have important roles to play.

Forest ecosystems provide many benefits that we humans depend on. They form the headwaters of major river systems. They sustain biological diversity and wildlife habitats in extraordinarily rich ecosystems (Wilson 2002). They are, obviously, the source of wood, a raw material environmentally superior to alternatives. For example, comparing the energy, effluent, and emissions consequences of using wood and steel in constructing commercial buildings, wood outperforms steel in every category, often by an order of magnitude or more (Bowyer 2001).

Moreover, wood products meet many human needs, such as housing, furniture, and paper. Wood is still an important global energy source. In some developing countries, as much as 80% of the energy for cooking and heating in rural areas still comes from wood (Gardner-Outlaw and Engleman 1999). Wood and trees are also significant in carbon sequestration. Brown (1997) estimates that the amount of carbon stored in the world's forests is about equal to the amount of $CO_2$ in the atmosphere. Beyond that, wood in use for housing, furniture, and containers is still sequestering carbon. Most of the carbon stored in the world is in the oceans, but forests and wood products are the big players on land. Finally, besides embodying a multitude of recreational and spiritual values, forests are a major part of the identities of many cultures across the globe. For anglers, hunters, and hikers, forests are key aspects of their quality of life.

Forests now cover about a quarter of the land surface of the world (FAO 2002) and about 33% of the area of the United States. Along the U.S. Pacific coast, forest cover ranges from just under 40% in California to about 50% in Washington (USDA Forest Service 2001). Globally and in the United States, this is less forest than existed 400 years ago because people have transformed forests through a wide variety of activities, the most significant of which has been the conversion of forests into agriculture and human dwellings (Perlin 1991; Williams 1989; MacCleery 1996; FAO 2002).

Management activities also transform forests. Even when they do not change them from forests to something else, they do change their character, structure, and species composition. Examples of changes through management and other human activities are livestock grazing, recreation, human-induced climate change, roads, water diversions, and the harvest and replanting of trees. Thousands of hectares of former riparian forests now also sit beneath water reservoirs in the United States alone. World population growth is the major force behind forest transformation, and we humans will increase our numbers by one-third to one-half by midcentury (Gardner-Outlaw and Engleman 1999).

Some forest transformations tend to be relatively permanent, at least when thought of in human lifetimes, but of course they are not permanent over geological time. Over very long periods, some of the imprints of humanity will be erased. In general, however, changes such as urban sprawl and agricultural transformations are considered to be forest lost. Not all transformations are losses; people also plant forests, sometimes after tree harvest and sometimes on abandoned agricultural lands, transformations that are usually considered net benefits to the environment. Even so, global forest losses are outstripping gains, with the loss of 12 million hectares annually in tropical countries only partially offset by the gain of 2.8 million hectares per year in temperate nations (FAO 2002). If the losses are not reversed and population growth continues, it is inevitable that a smaller forested area will have to serve more people in the future. As those people gain in knowledge and affluence, it is likely they will demand that forests serve them in more ways. In addition to water and wood, biodiversity protection, recreation, carbon sequestration, and personal solitude will become increasingly valuable forest uses and services.

## Managed Forests for Future Needs

This might sound like a gloomy future, but there is also good news for forests. We appear to be within a decade or two of a point at which approximately 40% of the industrial wood fiber produced in the world will come from planted forests (Victor and Ausubel 2000). It is the hope and expectation of many foresters and conservationists that planted forests will relieve some of the pressure to further develop native forests for wood production. Obtaining such environmental gains from planted forests, however, as Friedman and Charnley describe further on in this book, requires active public or private interventions.

Without management to sustain forest values to owners, experience has shown that people convert forests to nonforest uses, such as agriculture, urban growth, or livestock grazing, or they simply harvest too much and do not reinvest in the future forest, with consequences for the soils, plants, forage, wildlife, and water. With sustainable management, however, as foresters have learned in many parts of the world, forests can be restored and protected and their productivity for wood enhanced. Although management cannot quickly restore the same kind of native forest that existed before human intervention, it can sustain and protect diverse, productive forests. Let us consider what it will take to sustain forests for all their values and uses.

## Forest Sustainability

Sustainability is a concept in which the environmental, economic, and social dimensions of interactions between people and resources are equally important. It is an idea that is now shaping perspectives on forest management in most nations (FAO 2002). Sustainability requires meeting both current and future needs, and it thus implies intergenerational equity (WCED 1987). It seeks a balance among environmental protection, economic development, and the perpetuation of vibrant human communities. It views people as part of nature, part of ecosystems, not as a separate entity to be dealt with after figuring out how to take care of the non-people parts. Because human uses of forest resources now involve global wood markets, forest sustainability also means we must consider regional, national, and global forests in our local decisions.

Local forest sustainability rests in a global context; we cannot accomplish it by focusing only on local matters, or even regional or national ones. Let me illustrate: Large volumes of wood move across international boundaries every year in the global wood trade, about one-third of all industrial wood consumed each year. Southern hemisphere countries are major suppliers to northern hemisphere countries. Forest-related businesses, such as International Paper and Weyerhaeuser in the United States and UPM-Kymmene and Stora-Enso of Finland and Sweden, started out as local or regional companies, then expanded to become national corporations, and now are globally integrated corporations, with lands and mills on different continents. They market their products in a global marketplace.

Because forests are important for many different values, uses, products, and services, they are managed for many different outcomes—some to produce materials, some to provide recreational opportunities, and some to protect natural habitats. Sustainable forestry must therefore be as broad as the many different purposes for forests that people wish to sustain. It cannot be defined by a single approach that can be applied to the management and protection of every different kind of forest. It must have variants for national parks, for multiple-use forests, and for forests where wood production is the primary purpose. Across this spectrum we can consider three major types of sustainable forests: wood production forests, nature or reserve forests, and integrated multiresource forests.

### Three Kinds of Sustainable Forestry

Most of the world's industrial wood eventually will come from forests planted specifically for wood production (Sedjo and Botkin 1997). As mentioned above, we are already well on the way in our transition from extracting much of our wood from natural or seminatural forests to growing and harvesting most of it from forests planted for and dedicated to producing high yields of wood.

There is much potential for applications of biotechnology and genetically modified organisms in wood production forestry (Sedjo 2001; Wilson 2002). One obvious application will be to increase productivity. Equally important, however, are applications that could reduce environmental impacts of intensive management, for example, by reducing the volume and toxicity of fertilizer, pesticide, and other chemical treatments and perhaps even reducing water requirements of production, while improving wood quality and consistency. If predictions by leading scientists are correct, eventually only 10 to 15%—perhaps even less—of the world's forested area will be in forests planted and managed for high yields of wood, meeting more than 80% of the world's industrial wood needs by 2050 (Sedjo and Botkin 1997; Victor and Ausubel 2000).

The second major type of sustainable forestry—we might call it nature preservation or reserve forestry—is commonly practiced in areas such as national parks, wildlife reserves, and wilderness areas. These forests sustain an array of forest values and services not possible in forests managed for moderate to high levels of products or human uses. It is common to think that parks, wildlife reserves, and wilderness areas are not managed. They are, but the focus of management is not on forest products.

In the United States, reserved forests are managed to limit human activities so as to minimize their impact on native species and ecosystems and to prevent nonnative, invasive species from displacing natives. Some U.S. national parks have reintroduced managed fire, for example, and even commercial timber sales, to thin smaller trees that create fire hazards to large, old trees and return forest structure to conditions that can withstand natural disturbances.

Reserve forests are not managed for economic gain, but even here there may be some potential for applications of biotechnology in restoration of native species endangered by exotic diseases and pests. Biotechnology could also indirectly benefit reserve forests by enhancing the ability of wood production forests to meet most of the world's wood needs. Victor and Ausubel (2000) suggest that use of planted forests to meet wood needs could allow as much as 50% of world forests to remain in undeveloped status as reserves.

The remaining forests—the majority of the world's accessible forests—will likely be in some intermediate kind of stewardship or management for multiple benefits (Sedjo and Botkin 1997; Gardner-Outlaw and Engleman 1999; Dudley et al. 1996). Biotechnology might also help here in optimizing joint production of the multiple benefits. A major role of multiresource forests will be to protect endangered and at-risk species and ecosystems where some level of wood production is desired. If we can learn how to efficiently map and move resistance genes into vulnerable strains of native trees, we may also be able to restore and sustain trees that have been severely affected by exotic pathogens.

## *Different Roles for Different Owners*

Ownership matters in deciding which approach to sustainable forestry to apply. Much of the public focus in the past two decades in the United States has been on federal and other public forests, which make up only 42% of America's forestland. Of the public forests, 16% is reserved by congressional or state actions as parks, refuges, or wilderness, and 38% is uneconomical for wood production or is in a land use that precludes tree harvest. This leaves 46% of public forest, only 19% of total U.S. forestland, available to contribute to the nation's wood needs through either wood production or multiresource approaches to sustainable forestry.

The most productive and largest area of forestland in the United States is privately owned (USDA Forest Service 2001). Industrial ownerships account for 9% of total U.S. forestlands (13% of total commercial timberland). Small to modest-sized tracts of land in family ownerships comprise 49% of U.S. forests (58% of total commercial timberland).

Clearly ownership has a major influence on roles in regional and national forest sustainability. National forests and national parks will dominate the nature preservation, reserve forest end of the spectrum. Industry forests will emphasize wood production. Family-owned forests, due to diverse owner preferences, will encompass all approaches to sustainable forestry. (The ownership of forestlands in other countries varies considerably, ranging from almost entirely public to almost entirely private.)

# Sustainability: More Than Forest Management

Although forest management can determine where and how wood will be produced and where forests will be dedicated to other values, forest management alone is not sufficient for all that societies aspire to sustain. Demand for forest uses, products, and benefits—whether water, recreation, wood, or biodiversity—comes from everyone who uses or wants those products or services. Achieving forest sustainability will require that all people, including forest managers, forest products manufacturers, and the users of forests and their products, make intelligent choices in what they demand and how they use resources. What we choose to use, and how; where we decide to produce it and through what technologies; and what we decide to do with it when we are through are all important to sustainability.

We are moving into a future in which more people will demand more from forests whose area and availability to meet resource needs are likely to be shrinking. Those human needs and demands for forest values, uses, products, and services can only be met through a combination of tempering some of the demands and increasing our ability to produce what is needed from those forests where production is most efficient and beneficial to the entire environment. This means that societies must continually invest in new knowledge and technologies, so as to learn how to better integrate protection of water, fish, and wildlife with increased wood production in planted forests. Societies will need to invest in learning how to provide the multiple benefits of other managed forests and how to perpetuate the natural values desired from reserve forests. It is probable that biotechnology applied to these purposes will eventually be a necessary tool in the full spectrum of sustainable forestry.

## *A Strategy Framework: Progress, Responsibility, and Common Ground*

Here I take a bit of a risk and offer a framework to organize our thinking and our dialogue on forest sustainability and roles for biotechnology. I do not suggest that this is a complete set of necessary and sufficient steps—or that any of the steps will be easy.

- Design ecosystem transformations and biotechnology applications such that overall sustainability—its social, economic, and environmental dimensions—is enhanced. We cannot stop ecosystem transformations or biotechnology developments in a growing, technologically inquisitive population, but we can make them more conducive to environmental health, economic vitality, and community livability.
- Restore a conservation ethic that emphasizes prudent use of renewable natural resources.
- Emphasize the production and use of solar-powered renewable resources, produced with as little environmental impact as possible and with sensi-

tivity to social equity in how the benefits and costs of production are distributed locally, nationally, and globally. This is especially important for the United States, as we are the largest consumer of natural resources, renewable and nonrenewable, and we shift much of the economic, social, and environmental impact of our consumption to nations and cultures less able than ours to handle them.

- Develop and apply knowledge, technology, and governance systems to sustain desired social, environmental, and economic conditions, approaching these desired conditions simultaneously. This is enormously difficult to do; humans tend to pursue their goals one at a time. We have economic development schemes, and then we have regulatory schemes to stop them, to protect the environment. Often in both schemes, people, cultures, and communities become lost. The social dimensions of sustainability are paramount and cannot be afterthoughts.
- Manage ecosystems and technologies—especially the human enterprises intertwined with them—to meet these combined social, environmental, and economic goals.
- Include equity in the aspiration for sustainability. Not everyone is benefiting at the same rate and to the same degree in the economic development and environmental improvements of this world. Tolerating a widening gap between affluent and poverty-stricken people and cultures is not a sustainable proposition.

It is inevitable that technological advances will occur somewhere in the world to help meet human needs; they always have, and there is no reason to think the process will stop. If those advances do not occur in places such as universities and the open, democratic, and rigorous scientific forum from which this book originated, then they will occur behind closed doors, without safeguards, and without public dialogue. We have here not a question about whether biotechnology or genetically modified organisms will emerge and get into the environment. Rather, it is a question of who is going to set the ground rules for how that is done and to what degree we will use this technology in managing nature and in equitably meeting human needs.

## Defining Sustainable Forest Biotechnology

Among the concerns about transgenic trees that this book addresses are the following:

- Will transgenic trees enhance productivity, and if so, to what degree?
- Will they help reduce the environmental impact of meeting people's needs for resources, and if so, to what degree?
- Can they be used to mitigate some of the undesired consequences of global climate change?

- How should we deal with intellectual property rights in connection with genetically modified trees?
- What are the ethical dimensions of intervening with creation or nature?
- How are people to make decisions about all of this? Who will make the decisions? Who is going to be participating, and who might be left out?

In many parts of the world, people are squabbling over forests, taking polar, "all-or-none" positions and relegating various technologies to one side or the other, whether for technical or symbolic reasons. Extremist ideologies abound. They feed on controversy, and where none exists, they create it. There are people who think that markets know best: Just turn everything over to the marketplace and everything will be fine. Some assert that scientists know the answers: Let's ask the scientists and let them tell us what to do. Some people think that nature knows best: Just leave nature alone and let it do its thing. Such ideologies taken to the extreme destroy any potential for common ground and mutual progress. They leave no room for reasonable and responsible people to consider and weigh the multiple consequences of the hard choices required to improve the human condition.

It is time to open a dialogue about the consequences of our choices on sustainability and roles for biotechnology. If genetically engineered trees are not to be part of our path to forest sustainability, then how are we to meet the wood and other forest benefit needs of eight to ten billion people? If transgenic trees are to be part of that path, then what rules will guide their use? The choices are up to us, and all of them have consequences.

Our common future is a world with more people and more pressures on finite lands and natural resources. We are also on a trend of widening gaps between affluent and poverty-stricken cultures. We must meet people's basic needs for food, water, clothing, shelter, and dignity and provide a quality environment that continues to perform its fundamental ecosystem services. This can only be done by gaining efficiencies in the production and distribution of basic resources, so as to relieve some natural systems from human pressures.

Open, democratic societies are the most likely places for biotechnology developments to occur safely and with maximum human benefit. If those of us in such countries default because we cannot agree on how to proceed, the technologies will be developed in places where we will be much less certain about safety and beneficial effects. That would be irresponsibility of the highest magnitude.

## References

Bowyer, J.L. 2001. Environmental Implications of Wood Production in Intensively Managed Plantations. *Wood and Fiber Science* 33(3): 318–333.
Brown, S. 1997. Forests and Climate Change: The Role of Forest Lands as Carbon Sinks. In *Proceedings of the XI World Forestry Congress, Anatalya, Turkey.* Rome, Italy: U.N. Food and Agriculture Organization.

Dudley, N., D. Gilmour, and J.-P. Jeanrenaud. 1996. *Forests for Life—The WWF/IUCN Forest Policy Book.* Godalming, Surrey, UK: WWF-UK, Panda House.

FAO (Food and Agriculture Organization of the United Nations). 2002. *Global Forest Resources Assessment 2000.* FAO Forestry Paper 140. Rome, Italy: FAO.

Gardner-Outlaw, T., and R. Engleman. 1999. *Forest, Population, Consumption and Wood Resources.* Washington, DC: Population Action International.

MacCleery, D.W. 1996. *American Forests: A History of Resiliency and Recovery.* Durham, NC: Forest History Society.

Perlin, J. 1991. *A Forest Journey. The Role of Wood in the Development of Civilization.* Cambridge, MA: Harvard University Press.

Sedjo, R. 2001. Biotechnology's Potential Contribution to Global Wood Supply and Forest Conservation. Discussion Paper 01–51. Washington, DC: Resources for the Future. http://www.rff.org/disc_papers/PDF_files/0151.pdf (accessed December 15, 2002).

Sedjo, R., and D. Botkin. 1997. Using Forest Plantations to Spare Natural Forests. *Environment* 39(10): 14–22.

USDA Forest Service. 2001. *U.S. Forest Facts and Historical Trends.* FS-696. Washington, DC: USDA Forest Service.

Victor, D.G., and J.H. Ausubel. 2000. Restoring the Forests. *Foreign Affairs* 79(6): 239–257.

WCED (World Commission on Environment and Development). 1987. *Our Common Future.* New York: Oxford University Press.

Williams, M. 1989. *Americans and Their Forests: A Historical Geography.* New York: Cambridge University Press.

Wilson, E.O. 2002. *The Future of Life.* New York: Knopf.

# 2

# Biotechnology and the Forest Products Industry

### ALAN A. LUCIER, MAUD HINCHEE,
### AND REX B. MCCULLOUGH

**B**oth large and complex, the forest products industry employs millions of people with diverse skills all over the world (Diesen 1998; Smith 1997). Among the many human needs the industry serves are housing, furniture, packaging, communication, artistic expression, recreation, personal hygiene, energy, and chemicals (see Table 2-1).

There are good reasons for optimism about the future of the forest products industry. World demand for forest products should increase substantially with increases in population and greater prosperity (Diesen 1998; FAO 2001; Haynes 2003; Perez-Garcia 2001). Wood also has inherent environmental advantages relative to other raw materials.

- Demand for wood can provide important incentives for afforestation, reforestation, and sustainable forest management (Adams et al. 1996) and thus can have positive effects on the extent and condition of forest resources.
- Most forest products are renewable and recyclable. Moreover, product life cycle studies show that forest products typically require less fossil energy during manufacture and use than competing materials (NRC 1976). For example, a recent life cycle study of design options for a typical single-family house in Minnesota found that a steel design required 1.7 times the energy of a wood design, with a net difference in greenhouse gas emissions equivalent to 18.5 metric tons of carbon dioxide per house (Bowyer et al. 2001).
- Residuals from wood processing are important sources of renewable energy (EIA 2001). The forest products industry is already the world

**TABLE 2-1.** Examples of Forest Products That Help Meet Important Human Needs

*Forest resources*
— sawlogs, pulpwood, fuel wood
— recreation opportunities
— ecosystem services such as water purification, carbon sequestration, wildlife
  habitat

*Building materials*
— framing lumber, structural panels, siding, beams, floor joists, roof trusses, interior
  paneling
— paper components of wall board, counters, and insulation

*Communication papers*
— books, newspapers, magazines
— office papers, stationery, school and note pads, drawing paper
— greeting cards, poster, and display boards

*Packaging*
— boxes, bags, drums, tubes, spools, cores
— paperboard for food packaging, milk cartons, juice cartons
— pallets, wood shipping containers

*Tissue and absorbent fibers*
— personal hygiene products
— paper towels
— diapers
— convalescent bed pads

*Specialty cellulose*
— acetate textile fibers
— photographic films
— plastics, pharmaceuticals, food products
— thickeners for oil drilling muds
— rayon for tires and industrial hoses

*Other forest products*
— steam and electricity from biomass fuels
— Christmas trees
— envelopes, labels, file folders
— toys, decorations, sporting goods
— mulch, compost, wood ash, and other soil amendments
— railroad ties, utility poles
— landscaping timbers, fence posts
— disposable cups and plates, take-out food containers
— furniture, tool handles, musical instruments
— specialty chemicals, fragrances

leader in biomass energy production and will probably increase its production substantially if new technologies such as biomass gasification are successful (Tucker 2002). Greater use of woody biomass to produce energy could create significant new demand.

Optimism about the industry's future is tempered by serious challenges. The land base for future wood production will be constrained severely by competing land uses, such as agriculture, residential development, and wilderness preservation (FAO 2001; Sedjo 2001; Sedjo and Botkin 1997; Victor and Ausubel 2000). In addition, the industry is contending with dynamic and difficult market conditions, a large and growing number of government regulations with major impact on the industry, and important stakeholder initiatives such as forest certification (Diverio 1999; IIED 1995; Lucier and Shepard 1997; Rooks 2000; Smith 1997; Swann 1999).

The industry's wood supply challenge can be overcome by increasing production on lands well suited to intensive silviculture and by developing landscape management strategies that improve the overall condition of forest ecosystems (Lucier 1994; Lucier et al. 2002; Sedjo and Botkin 1997). Forest managers are making substantial progress in these directions using technologies as diverse as genetic tree improvement, weed control, wildlife management, and landscape design.

Manufacturing facilities in the forest products industry operate in many different countries and climates and are technologically diverse, encompassing lumber mills, several kinds of pulp and paper mills, and plants that make a wide variety of composite materials used in construction and many other applications. Strategic challenges to the industry globally include high capital costs for manufacturing facilities, low commodity prices, and a complex array of environmental and energy issues (AF&PA 2000; IIED 1995; Smith 1997). Among the long-term solutions the industry is exploring are new product development, workforce training, and innovations in manufacturing and environmental controls (TAPPI 2002).

Biotechnology has significant potential to help the forest products industry overcome these challenges. In this chapter, we describe some of the potential applications. We also discuss obstacles to progress in forest biotechnology generally and some promising paths forward.

## Biotechnology and Tree Improvement

The productivity and quality of agricultural crops have been greatly improved by centuries of breeding, testing, and genetic selection. Modern crop varieties are much better sources of food and fiber than their wild ancestors, and they greatly reduce the amount of land that must be cultivated to meet human needs (Borlaug 1997).

In comparison to agricultural crops, trees planted for wood production are wild, undomesticated plants. Most efforts to improve trees for wood production have been under way for less than 50 years. Results with several species (e.g., loblolly pine and Douglas fir) are scientifically and economically important. They confirm expectations of plant breeders, based on genetic theory and agricultural experience, that tree species have enormous genetic potential that could be expressed in valuable new varieties (Burley 2001; Zobel and Talbert 1984). For example, loblolly pine varieties with fast growth and superior wood properties would be economically valuable to both tree growers and forest products manufacturers.

Progress in forest tree improvement has been constrained by various difficulties inherent in tree breeding and propagation (Burdon 1994; Huang et al. 1993). For example, many important species are "self-incompatible," meaning that viable progeny cannot be produced by combining male and female gametes from the same tree. This prevents establishment of inbred lines, a technique that enables rapid genetic improvement through selection and breeding in self-compatible species such as corn.

Tree breeding typically involves crossing many individual parent trees and testing the progeny of the breeding pairs. Progress is slow because the progeny tests must be monitored for 5 to 10 years or more to identify the best parents and progeny with confidence. Parents that consistently produce superior progeny are crossed with each other on a large scale in production seed orchards. The best progeny of the best parents are selected and established in new breeding orchards. Several years after establishment, trees in the new breeding orchard will begin to flower, and a new cycle of crossing and progeny testing can begin.

In many plant species (e.g., poplars and grapes), there are efficient methods for multiplying the best individual genotypes using vegetative propagation techniques such as induction of root formation on cut stem segments (the "rooted cutting method"). An efficient vegetative propagation system allows growers to quickly capture the benefits of selection and breeding by multiplying and planting the best genotypes on a commercial scale. In contrast, many important tree species are more difficult to propagate vegetatively, and it has not been economically feasible to multiply the best genotypes for planting in large-scale forestry operations. Instead, the best parents are crossed with each other to produce seeds and seedlings as we have just described. The resulting seedling families are quite diverse genetically, with average growth and wood quality considerably less than that of the best individuals.

Biotechnology has great potential to accelerate genetic tree improvement and enable production of higher value raw materials for the forest products industry (Burdon 1994; Burley 2001; Huang et al. 1993; Raemdonck et al. 2001). Key technologies include (a) advanced breeding strategies based on marker-aided selection; (b) vegetative propagation based on somatic embryogenesis; and (c) rapid introduction of valuable traits into superior germplasm by genetic engineering. Marker-aided selection involves molecular genetic analysis of parents and progeny to focus selection on individuals that carry genes associated with desirable traits. Somatic embryogenesis involves in vitro culture of embryonic tree tissue to enable efficient propagation of superior genotypes. Genetic engineering involves insertion of one or more genes into cells of superior genotypes (transformation) and in vitro propagation of transformed plants (regeneration).

If biotech approaches to tree improvement are successful technically and commercially, significant benefits would be realized at several points along the value chain from forest to final consumer. The following scenarios illustrate the distribution of potential benefits.

- Vegetative propagation and genetic engineering enable substantial increases in tree growth rates on sites close to mills. When ready for harvest, these sites yield large numbers of uniform stems per hectare. High yields reduce harvesting costs and the area of forestland required to meet mill demands for raw material. Short haul distances to mills reduce log transportation costs. Efficiencies in harvesting and transportation reduce fossil fuel consumption and carbon dioxide emissions associated with raw material acquisition (Lucier et al. 2002).
- Vegetative propagation and genetic engineering enable development of tree varieties with special wood properties tailored to the requirements of manufacturing processes (Whetten and Sederoff 1991; Hu et al. 1999). For example, a smaller core of juvenile wood would create greater lumber strength and stability. Higher specific gravity would mean higher pulp yields relative to inputs of energy and chemicals at the pulp mill; and lower lignin content would also reduce the chemicals and energy used in pulping and bleaching. Such improvements in raw material quality allow mills to reduce manufacturing costs and improve product quality.

Many other benefits from biotechnology and tree breeding are also possible (Mathews and Campbell 2000; Yanchuck 2001):

- New pest management strategies based on improvements in the genetic resistance of trees and reduced quantities of insecticides and fungicides (Punjar 2001; Strauss et al. 1991)
- Ecological restoration strategies enabled by genetic engineering of tree species that have been devastated by exotic diseases (e.g., American chestnut) (Carraway and Merkle 1997; IFB 2002)
- Carbon sequestration, soil reclamation, and bioremediation strategies using new tree varieties capable of tolerating poor soil conditions such as drought and chemical contamination (Pullman et al. 1998)
- New strategies for sustainable production of valuable chemicals in trees, based on genetic engineering of secondary metabolic pathways (Obst 1998)

Tree selection and breeding will continue to be the essential foundations of tree improvement whether or not the potential benefits of forest biotechnology are realized. Superior genotypes from selection and breeding programs are prerequisite for successful vegetative propagation and for genetic engineering. It would be unwise to view biotechnologies as alternatives to traditional methods of tree improvement. Rather, we consider the new tools of biotechnology important supplements to traditional methods.

## Biotechnology and Forest Products Manufacturing

The forest products industry is under great financial pressure at present (Diverio 1999; Swann 1999). Disappointing returns to shareholders, along with globalization, are driving a dramatic restructuring of the industry (Diesen 1998; Rooks 2000).

Although financial and market issues are dominant near-term concerns, many industry leaders have keen interest in the potential of technology to reduce manufacturing costs and create new products (AF&PA 1994, TAPPI 2002). Besides improving the quality and quantity of raw material supplies, over the long term biotechnology could have a radical impact on pulping processes, waste-to-energy systems, and other aspects of forest products manufacturing (Ericksson 1997; Luke 1994).

- Biotechnology could enable significant progress in the development of biopulping and biobleaching processes (Akhtar et al. 1998; Cullen 1997; Ericksson 1997), permitting lower capital costs, higher product quality, and reduced use of both chemicals and energy.
- Biotechnology could enable the development of new systems for converting organic residuals into bioenergy (Jeffries and Schartman 1999), resulting in lower costs for solid waste management and reduced need for fossil energy.

## Realizing the Potential of Forest Biotechnology

Realizing the great potential of biotechnology in the forest products industry is an enormous and exciting challenge. The rate of progress will depend not only on science and technology, but also on their interaction with social, economic, and political factors (Mathews and Campbell 2000; Tuskan 1998; Wolfenbarger and Phifer 2000).

Through its Agenda 2020 program, the forest products industry in the United States has suggested priorities for precompetitive research and provided funding for several biotechnology projects in partnership with the U.S. Department of Energy and the USDA Forest Service (Lucier et al. 2002; see Table 2-2).

**TABLE 2-2.** Examples of Forest Biotechnology Projects Supported through Agenda 2020

| Principal investigator | Project | Lead institution |
| --- | --- | --- |
| Brunner | Dominant negative mutations of floral genes for engineering sterility | Oregon State University |
| Chang | Exploiting genetic variation of fiber components and morphology in juvenile loblolly pine | North Carolina State University |
| Davis | Molecular physiology of nitrogen allocation in poplar | University of Florida |
| Davis | Molecular determinants of carbon sink strength in wood | University of Florida |
| Li | Search for major genes using progeny test data to accelerate development of superior loblolly pine plantations | North Carolina State University |
| Neale | Genetic marker and quantitative trait loci mapping for wood quality traits in loblolly pine and hybrid poplars | USDA Forest Service |
| Peter | Accelerated stem growth rates and improved fiber properties of loblolly pine | Institute of Paper Science and Technology |
| Pullman | Trees containing built-in pulping catalysts | Institute of Paper Science and Technology |
| Tsai | Genetic augmentation of syringyl lignin in low-lignin aspen trees | Michigan Technological University |
| Tschaplinski | Biochemical and molecular regulation of crown architecture | Oak Ridge National Lab |
| Tuskan | Marker-aided selection for wood properties in loblolly and hybrid poplar | Oak Ridge National Lab |
| Whetten | Pine gene discovery project | North Carolina State University |
| Williams | QTL and candidate genes for growth traits in *Pinus taeda* L. | Texas A&M University |

Agenda 2020 and other programs are supporting valuable projects, but meager overall government funding for precompetitive research critically limits forest biotechnology. Mapping the genomes of model tree species and discovering molecular controls of key processes such as wood formation are formidable tasks that will take decades at current rates of progress. A major initiative is needed to accelerate precompetitive research in these areas (Sederoff 1999).

The ecological, social, and policy issues associated with forest biotechnology are complex and extremely important (Mathews and Campbell 2000). Many of the key issues are discussed in detail elsewhere in this volume. In our view, greater understanding and more effective management of those issues are critical for realizing the potential benefits of forest biotechnology. Progress is most likely to be achieved through discussion, research, and collaboration involving all interested parties, including both supporters and critics of biotechnology and the forest products industry. A new Institute of Forest Biotechnology (www.forestbiotech.org) has been established through collaborative efforts of government, industry, and nonprofit organizations to serve as a convener of interested parties and to help organize the necessary exchanges of ideas.

## Conclusion

Forest biotechnology holds important opportunities and challenges for the forest products industry. The industry's technology leaders appreciate the economic potential of forest biotechnology and have diverse views on critical issues such as time to commercialization and the management of risk.

Here we have emphasized the value of forest biotechnology to the forest products industry. However, society at large could also benefit substantially through (a) new supplies of renewable energy and materials, (b) effective new options for solving difficult problems of environmental management and ecological restoration, and (c) new opportunities for employment and sustainable development in an industry based on renewable resources. The future of biotechnology and its value to the forest products industry will be affected greatly by public perceptions of its social and ecological ramifications. We believe the potential benefits of forest biotechnology that we have outlined in this chapter justify greater public support for precompetitive research to advance the science, to develop and test promising applications, to evaluate ecological risks and social concerns, and to develop appropriate public policies.

## Authors' Note

The authors thank Dawn Parks at ArborGen for helpful comments and suggestions.

# References

Adams, D.M., R.J. Alig, J.M. Callaway, B.A. McCarl, and S.M. Winnett. 1996. The Forest and Agricultural Sector Optimization Model (FASOM): Model Structure and Policy Applications. Research Paper PNW-RP. Portland, OR: USDA Forest Service, Pacific Northwest Research Station.

AF&PA (American Forest and Paper Association). 1994. *Agenda 2020: A Technology Vision for America's Forest, Wood, and Paper Industry.* Washington, DC: American Forest and Paper Association.

————. 2000. *AF&PA Environmental, Health and Safety Verification Program: Year 2002 Report.* Washington, DC: American Forest and Paper Association.

Akhtar, M., G.M. Scott, R.E. Swaney, and T.K. Kirk. 1998. Overview of Biomechanical and Biochemical Pulping Research. In: *Enzyme Applications in Fiber Processing. ACS Symposium Series* 687: 15-26. Washington, DC: American Chemical Society.

Borlaug, N.E. 1997. Feeding a World of 10 Billion People: The Miracle Ahead. *Plant Tissue Culture and Biotechnology* 3:119–127.

Bowyer, J., D. Briggs, L. Johnson, B. Kasal, B. Lippke, J. Meil, M. Milota, W. Trusty, C. West, J. Wilson, and P. Winistorfer. 2001. CORRIM: A Report of Progress and a Glimpse of the Future. *Forest Products Journal* 51(10): 10–22.

Burdon, R.D. 1994. The Role of Biotechnology in Forest Tree Breeding. *Forest Genetic Resources* 22: 2–5.

Burley, J. 2001. Genetics in Sustainable Forestry: The Challenges for Forest Genetics and Tree Breeding in the New Millennium. *Canadian Journal of Forestry Research* 31: 561–565.

Carraway, D.T., and S.A. Merkle. 1997. Plantlet Regeneration from Somatic Embryos of American Chestnut. *Canadian Journal of Forest Research* 27: 1805-1812.

Cullen, D. 1997. Recent Advances on the Molecular Genetics of Lignolytic Fungi. *Journal of Biotechnology* 53: 273–289.

Diesen, M. 1998. *Economics of the Pulp and Paper Industry.* Book 1 in the series Papermaking Science and Technology. Helsinki, Finland: Fapet Oy.

Diverio, M.X. 1999. Poor Financial Returns Are a Call to Action for the Paper Industry. *Pulp and Paper,* March, 71–81.

EIA (Energy Information Administration). 2001. *Renewable Energy Annual 2000.* DOE/EIA-0603(2000). Washington, DC: Energy Information Administration, U.S. Department of Energy.

Eriksson, K-E.L. 1997. *Biotechnology in the Pulp and Paper Industry.* New York: Springer-Verlag.

FAO (UN Food and Agriculture Organization). 2001. *Global Forest Resources Assessment 2000.* FAO Forestry Paper 140. Rome, Italy: United Nations Food and Agricultural Organization.

Haynes, R.W. 2003. *An Analysis of the Timber Situation in the United States: 1952–2050.* General Technical Report PNW-GTR-560. Portland, OR: USDA Forest Service, Pacific Northwest Research Station.

Hu, W.J., J. Lung, S.A. Harding, J.L. Popko, J. Ralph, D.D. Stokke, D.D. Tsai, and V.L. Chiang. 1999. Repression of Lignin Biosynthesis Promotes Cellulose Accumulation and Growth in Transgenic Trees. *Nature Biotechnology* 17: 808–812.

Huang, Y., D.F. Karnosky, and C.G. Tauer. 1993. Applications of Biotechnology and Molecular Genetics to Tree Improvement. *Journal of Arboriculture* 19(2): 84–98.

IFB (Institute of Forest Biotechnology). 2002. *Heritage Trees Workshop Meeting Report.* Research Triangle Park, NC: Institute of Forest Biotechnology.

IIED (International Institute for Environment and Development). 1995. *Towards a Sustainable Paper Cycle.* London: International Institute for Environment and Development.

Jeffries, T.W., and R. Schartman. 1999. Bioconversion of Secondary Fiber Fines to Ethanol Using Counter-current Enzymatic Saccharification and Co-fermentation. *Applied Biochemistry and Biotechnology.* 77–79: 435–444

Lucier, A.A. 1994. Criteria for Success in Managing Forested Landscapes. *Journal of Forestry* 92(7): 20–25.

Lucier, A.A., J. Pait, and P. Farnum. 2002. A "Green Goal": Sustainable Production of Higher Value Raw Materials. *Solutions,* June, 55–58.

Lucier, A.A., and J.P. Shepard. 1997. Certification and Regulation of Forestry Practices in the United States: Implications for Intensively Managed Plantations. *Biomass and Bioenergy* 13: 193–199.

Luke, D.L. 1994. Leadership, Choices and Technology. In: *TAPPI Proceedings 1994: Biological Sciences Symposium.* Atlanta, GA: TAPPI, 5–8.

Mathews, J.H., and M.M. Campbell. 2000. The Advantages and Disadvantages of the Application of Genetic Engineering to Forest Trees: A Discussion. *Forestry* 73(4): 372–380

NRC (National Research Council). 1976. *Renewable Resources for Industrial Materials.* Washington, DC: National Academy of Sciences.

Obst, J.R. 1998. Special (Secondary) Metabolites from Wood. In *Forest Products Biotechnology,* edited by A. Bruce and J.W. Palfreyman. London: Taylor and Francis, 151–165.

Perez-Garcia, J. 2001. Structural Changes in the Global Forest Sector in the 21st Century. *CINTRAFOR News* 18(4): 1,3. Center for International Trade in Forest Products, University of Washington, Seattle.

Pullman, G., J. Cairney, and G. Peter. 1998. Clonal Forestry and Genetic Engineering: Where We Stand, Future Prospects, and Potential Impacts on Mill Operations. *TAPPI Journal* 8: 57–64.

Punjar, Z.K. 2001. Genetic Engineering of Plants to Enhance Resistance to Fungal Pathogens: A Review of Progress and Future Prospects. *Canadian Journal of Plant Pathology* 23: 216–235.

Raemdonck, D.V., M. Jaziri, W. Boerjan, and M. Baucher. 2001. Advances in the Improvement of Forest Trees through Biotechnology. *Belgian Journal of Botany* 134(1): 64–78.

Rooks, A. 2000. Managing for the 21st Century. *PIMA's North American Papermaker* 82(1): 38–41.

Sederoff, R.R. 1999. Tree Genomes: What Will We Understand about Them by the Year 2020? In *Forest Genetics and Sustainability,* edited by C. Matyas. Dordrecht, Netherlands: Kluwer.

Sedjo, R. 2001. Biotechnology's Potential Contribution to Global Wood Supply and Forest Conservation. Discussion Paper 01-51. Washington, DC: Resources for the Future.

Sedjo, R., and D. Botkin. 1997. Using Forest Plantations to Spare Natural Forests. *Environment* 39(10): 15–20, 30.

Smith, M. 1997. *The U.S. Paper Industry and Sustainable Production: An Argument for Restructuring.* Cambridge, MA: MIT Press.

Strauss, S.H., G.T. Howe, and B. Goldfarb. 1991. Prospects for Genetic Engineering of Insect Resistance in Trees. *Forest Ecology and Management.* 43: 181–209.

Swann, C.E. 1999. Wall Street and the Paper Industry. *PIMA's North American Papermaker* 81(3): 36–39.

TAPPI. 2002. *Setting the Industry Technology Agenda: The 2001 Forest, Wood and Paper Industry Technology Summit.* Atlanta, GA: TAPPI Press.

Tucker, P. 2002. Fueling Self-sufficiency: Pathways to Change. *Solutions,* January, 67–70.

Tuskan, G.A. 1998. Short Rotation Woody Crop Supply Systems in the United States: What Do We Know and What Do We Need to Know? *Biomass and Bioenergy* 14(4): 307–315.

Victor, D.G., and J.H. Ausubel. 2000. Restoring the Forests. *Foreign Affairs* 79(6): 127–144.

Whetten, R., and R. Sederoff. 1991. Genetic Engineering of Wood. *Forest Ecology and Management* 43: 301–316.

Wolfenbarger, L.L., and P.R. Phifer. 2000. The Ecological Risks and Benefits of Genetically Engineered Plants. *Science* 290: 2088–2093.

Yanchuk, A.D. 2001. The Role and Implications of Biotechnological Tools in Forestry. *Unasylva* 52(1): 53–61.

Zobel, B.J., and J.T. Talbert. 1984. *Applied Forest Tree Improvement.* New York: Wiley.

# 3

# Biotechnology and the Global Wood Supply

### Roger A. Sedjo

Forestry is undergoing an important transition, as forests change from a wild resource, which had typically been foraged, to being a planted agricultural crop that is harvested periodically, like other agricultural commodities. The transition of forestry from foraging to agricultural cropping has been under way on a significant scale only within the past half-century or less (Sedjo 1999). Planted forests benefit from the same types of innovation that are common in other agriculture. When investors can capture the benefits of those improvements and innovations, economic incentives for investment in plant domestication, breeding, and plant improvement activities occur.

As in other types of agriculture, early tree improvements involved identification of trees with desired traits, sometimes called superior trees, and attempts to capture offspring that also had the desired traits. In the 1990s, however, forestry began to embrace modern biotechnology in earnest. A large number of applications (124) to field test transgenic trees have been filed in the United States, although to date only one transgenic tree species (papaya) has been deregulated and thus authorized for unrestricted commercial use in the country (McLean and Charest 2000). Canada too has ongoing transgenic tree activities (Mullin and Bertrand 1998).

The introduction of biotechnology to forestry has the potential to produce great economic benefits in the form of lower costs and increased availability to consumers of wood and wood products. Biotechnology can increase the productivity of planted forests and thus free large areas of natural forest from pressures to produce industrial wood. Genetically engineered trees may also be used to create novel products (e.g., plastics or pharmaceuticals), adding value to the more traditional wood and fiber products.

The domestication of a small number of plants, particularly wheat, rice, and maize, is among the most significant accomplishments of the human era. Modern civilization would be impossible without it. Plant domestication has brought high yields, large seeds, soft seed coats, nonshattering seed heads that prevent seed dispersal and thus facilitate harvesting, and a flowering time that is determined by planting date rather than by natural day length (Bradshaw 1999).

Recent decades have seen continuing increases in biological productivity, especially in agriculture, driven largely by technological innovations that have improved the genetics of domesticated plants and animals. Much of this improvement has been accomplished through traditional breeding techniques, in which desired characteristics of plants and animals, such as growth rates or disease resistance, can be incorporated into cultivated varieties.

The planting of genetically improved stock began about 1970 and accelerated in the 1990s. As more of the world's industrial wood is being produced in planted forests, the potential to introduce genetic alterations into the germplasm is obvious. Commercial forestry today may be on the threshold of the widespread introduction of biotechnology in the forms of sophisticated tissue cultures for cloning seedlings and genetically engineered trees.

## The Economics of Biotechnological Innovations in Forestry

Figure 3-1 is a simple schematic that illustrates the effects of the cost reductions that come from planted forests. In the absence of forest plantations, the volume of industrial wood harvested in a period is determined by the intersection of supply, $S$, and demand, $D$, at $e_0$. In this situation price is $P_0$ and the quantity harvested is $Q_0$. The introduction of relatively low-cost plantation forestry is represented by the line segment $aS'$. At price $P_1$ plantations provide a cheaper source of industrial wood than do natural forests. This new source of timber results in a new equilibrium, $e_1$, with a lower price, $P_1$, and a higher harvest volume, $Q_1$. Notice, however, that the volume harvested from natural forests is reduced from $Q_0$ to $Q_1'$. This reflects the fact that the low-cost plantation wood is displacing wood from natural forests. The effects of biotechnology are to reduce the costs of production, thereby shifting down even further the $aS'$ portion of the supply curve (not shown in Figure 3-1).

Genetic engineering can result in unique gene combinations not achievable by traditional tree breeding. Thus, in concept, frost-resistance genes could be transferred from plants or other organisms found in cold, northerly regions to tropical plants, thereby increasing their ability to survive in cooler climates.

Tree attributes or traits can be characterized as (a) silvicultural, (b) related to adaptability, and (c) related to wood quality (see Table 3-1). The

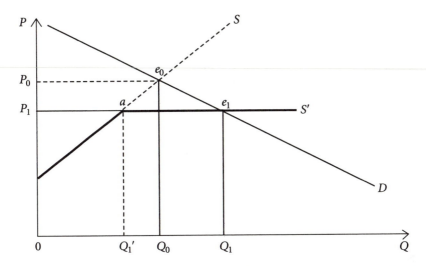

**FIGURE 3-1.** Schematic diagram of how increased production ($Q$) due to biotechnology or other sources can affect prices ($P$) and final harvest volumes. See text for definitions of other symbols.

silvicultural trait growth rate, for example, has a substantial genetic component, with rates differing as much as 50% between families or different clonal lines.

Yanchuk (2001) points out that biotechnologies used in forestry fall into three main areas: (a) the use of vegetative reproduction methods, (b) the use of genetic markers, and (c) the production of genetically engineered (GE) organisms, or transgenic trees. The first two generally involve applications over essentially the same time period as conventional breeding and thus need not involve delayed benefits. However, the use of GE techniques to produce benefits that are captured at harvest raises questions about the economic efficacy of that approach. Below I discuss some of these applications to forestry and examine timing and its implications for economic return.

## Economic Benefits

The economic benefits of biotechnology in forestry may include lower production costs and the development of new, higher quality products. The benefits may accrue to biotechnology developers and product producers in the short term, but through competition, in the longer term the benefits can be expected to accrue largely to consumers of the products, typically in the form of lower prices.

A distinguishing feature of the introduction of technology is increased productivity, or output per unit of input. Stated another way, one can think

**TABLE 3-1.** Tree Traits That Can Be Improved through Biotechnology

| Silviculture | Adaptability | Wood quality traits |
|---|---|---|
| Growth rate | Drought tolerance | Wood density |
| Nutrient uptake | Cold tolerance | Lignin reduction |
| Crown and stem form | Fungal resistance | Lignin extraction |
| Flowering control | Insect resistance | Juvenile fiber |
| Herbicide resistance | | Branching |

*Source:* Context Consulting provided information on potential innovations and their likely cost implications based on the best judgment of a panel of experts.

of technology as either cost reducing or yield (output) enhancing. This implies that society gets more output for its expenditure of inputs, and for the consumer, the relative price of desired goods falls compared with what it would have been in the absence of the innovation. Plantation forestry has enjoyed success in recent decades in part because it has generated cost reductions that give planted forests a competitive advantage over naturally regenerated or old-growth forests (Sedjo 1999).

In industrial wood production, there is an inherent incentive to improve the quality of the germplasm to generate tree improvements that can be captured at harvest. Tree improvements can take many forms (Table 3-2), but thus far, the most common emphases of tree improvement programs have been increased growth rates, improved stem form, and pest resistance. Disease and insect resistance traits, which promote or ensure the growth of the tree, may be desired to address specific problems common to particular species or to extend a species' climatic range.

For example, the development of frost-resistant eucalyptus would allow a much broader planting range for this commercially important genus. Other improvement possibilities include (as in agriculture generally) the introduction of an herbicide-resistance gene to permit more efficient use of effective herbicides, especially in the establishment phases of a planted forest. Besides promoting establishment, survival, and rapid growth of raw wood material, tree improvement programs can also focus on wood quality, including such characteristics as tree form, fiber quality, amount of lignin, and improved lignin extractability.

Desired output traits vary depending on the type of wood product for which the trees are intended. Wood quality improvement will seek to modify fiber characteristics in specific ways for pulping and paper production but seek entirely different kinds of wood quality changes for milling and carpentry products. Wood desired for furniture is different from that desired for framing lumber.

For paper, certain characteristics facilitate wood handling in the early stages of pulp production. For example, the straightness of the trunk has

**TABLE 3-2.** Possible Financial Gains from Future Biotech Innovations

| Innovation | Additional benefits[a] | Operating costs |
|---|---|---|
| Cloning of superior pine | 20% yield increase after 20 years | $40/acre or 15–20% increase |
| Wood density gene | Improved lumber strength | None |
| Herbicide-tolerance gene in eucalyptus (Brazil) | Reduced herbicide and weeding costs: potential savings of $350 or 45% per ha | None |
| Improved fiber characteristics | Reduced digester cost: potential savings of $10 per m$^3$ | None |
| Reduced amount of juvenile wood | Increased value: $15 per m$^3$ (more usable wood) | None |
| Reduced lignin | Reduced pulping costs: potential of $15 per m$^3$ | None |

[a]The actual cost savings experienced by the tree planter will depend on the pricing strategy used by the gene developer, the portion of the savings captured by the developer, and the portion passed on to the grower.

*Source*: Context Consulting.

value for improving pulp and paper products in that the lower amount of compression wood in straight trees generates preferred kinds of fibers. Straight trees are also more easily handled and fed into the production system. Ease in processing includes the breakdown of wood fibers in processing and the removal of lignin. Paper production requires fiber with sufficient strength to allow paper sheets to be produced on high-speed machines. The surface texture and brightness desired in some paper products are also properties that relate in part to the nature of the wood fiber used.

In addition, some characteristics are valued not for their utility in the final product but because they facilitate the production process. The absence of large or excessive branching, for example, reduces the size and incidence of knots, thereby allowing the use of more of the tree's wood volume. Straight trunks facilitate production of boards and veneer, and wood characteristics related to milling and use in carpentry include wood color, strength, and surface characteristics. Structural products such as oriented-strand board, fiberboard, and engineered wood products have their own unique sets of desired fiber characteristics.

Customized products require customized raw materials, and in recent years pulp producers have begun to move away from producing standardized "commodity" pulp into the production of specialized pulp for targeted markets. For example, Aracruz, a Brazilian pulp company, has asserted that it can customize its tree fibers to the requirements of individual customers. This requires increased control over the mix and types of wood fibers used. Thus far the control has been achieved through cloning but not transgenic plants.

## Development and Commercialization of Transgenic Trees

The development and commercialization of transgenics involve at least two components: first, the development of the transgene, and second, its introduction into the organism in sufficient numbers and at low enough cost. In many cases the transgene may be developed outside the forest company. In that case the developer, wishing to see a return on his investment, may require payment for the right to use the modified gene. The second component involves integrating the transgenic plants into the forest firm's nursery and planting system. Alternatively, forest planters might purchase genetically engineered seedlings directly from the developer or from an intermediate seedling producer.

Timing plays an important role in economic decisions about whether to use a modified gene in transgenic trees. The commercialization of transgenics typically requires investment in the early periods of germplasm development, returns on which are often delayed many years. Investments that require long periods before they produce a profit normally require bigger returns to compensate the investor for the period during which capital is tied up but not generating returns. Two common methods for assessing the economic viability of projects with large initial costs and delayed benefits are discounted present value and internal rate of return criteria.

Many forest operations take a long time to generate a profit. For example, an investment today in tree planting requires decades before a return is produced in harvested value. For that reason, the use of altered genes in certain forestry applications, though technically feasible, may be problematic. Nevertheless, it is clear that some long-term forestry projects deliver sufficient returns to justify the costs, including the opportunity costs over the long waiting period (Sedjo 1983), as is shown by the widespread and continuing establishment of plantation forests by profit-making firms.

A case where the benefits of a transgenic innovation may be captured quickly is the introduction of herbicide tolerance into trees. Broad-spectrum herbicides can then be applied to young forests to control weeds without injuring the trees, lowering establishment costs below those of traditional methods. In this application, genetic engineering allows the substitution of a low-cost weed-control technology at the beginning of the timber rotation (the cycle from planting to harvest). The benefits thus are "collected" in the form of cost savings at the beginning of the rotation, and no long waiting period is involved.

The quick capture of the economic benefits of herbicide-tolerant plants can be contrasted with the long delayed capture of economic benefits for fiber and lignin modification. For lignin, the costs of genetic engineering are incurred at the beginning of the rotation, but the benefits await the timber harvest and the lower costs experienced in the pulping process. However, the industry considers controlling lignin, a compound in trees that must be removed in the pulp-making process, an important priority. If the initial costs are sufficiently low and/or the subsequent benefits great, the innovation may generate acceptable economic returns despite the delay in capture. Aracruz Cellulose, in Brazil, has already undertaken trials of low-lignin trees developed through traditional breeding techniques, showing that the benefits and acceptance of risks from this kind of breeding are not unique to transgenics.

Other genetic alterations, such as flowering control, which delays flower initiation by several years, may be useful in preventing transgenic plants from transmitting modified genes to other plants and migrating into the wild. Such alterations may be prerequisite for the commercial release of some or all types of transgenic trees and thus can be viewed as an initial cost of doing this type of business. As with herbicide resistance, disease resistance may substitute for traditional disease control approaches in forestry. Economic viability depends on how effectively disease-resistant the transgenic trees are and how the costs compare with the traditional approaches. Finally, transgenic trees are likely to embody a number of altered genes. In an approach called "stacking," several desired genes are combined into single genotypes. Achieving acceptable economic returns will involve examining different possible gene packages in terms of their costs and benefits.

### Anticipated Cost Saving Innovations

A recent study identified several innovations in forest biotechnology believed to be feasible within the next decade or two and estimated the possible financial benefits of their introduction (Table 3-2). The calculations did not include intellectual property considerations (which can be complex) or the costs of developing the innovations. However, once the investment in innovation has been made, it is a fixed cost and unrelated to the marginal cost of the distribution of the product. The innovations suggest a potential decrease in costs and/or an increase in wood volume or quality, and approximate rates of return have been estimated for many of them.

For example, the 20% increase in volume due to the cloning of superior pine is estimated to provide a financial return of about 15–20% on the incremental investment cost of $40 per acre. This assumes initial yields of 15 m$^3$ per hectare (ha) per year and a stumpage price of $20 per m$^3$. In another example, in a Brazilian planted forest the herbicide-tolerance trait is estimated to generate an immediate saving in weed management costs of $350 per ha over the two to three years it takes to establish the plantation.

The benefits of biotechnological innovations that modify wood fiber characteristics to reduce pulping costs have also been estimated. The value added from pulping is about \$60 per m³, or \$275 per ton of pulp output. If these costs are reduced \$10 per m³, a surplus (or effective cost reduction) of about \$47 per ton of wood pulp (assuming 4.7 cubic meters per ton of pulp) would result, assuming wood prices are constant. This type of innovation would be important to the forestry sector because a mill would be willing to pay a premium to the producer for low-processing-cost wood fiber. If the improved fiber were to become common, the substantial processing cost savings could eventually be passed on to the consumer.

## The Global Impact: A Case Study of Herbicide Resistance

The potential costs savings for the future timber supply from one biotechnological innovation—the introduction of an herbicide-resistance gene—can be glimpsed in a case study on the costs of establishing commercial forests. By inference, the likely effect on harvests from natural forests is also examined. Here I take a crude partial equilibrium approach to estimate the cost savings associated with the application of the herbicide-resistance transgene in forestry. (A more sophisticated approach would involve integrating estimates into a forest sector systems model such as that of Sohngen et al. 1999.) The savings in plantation establishment costs are estimated on the basis of the data presented above, then translated into the lowering of the supply curve for planting activity. This results in an incremental addition to plantings. Because of the delay between planting and harvest, the direct impact on timber supplies is also delayed. However, the anticipation of greater future supplies, which would be expected to affect future prices, will affect current actions, including current harvests (Sohngen et al. 1999).

Figure 3-2 provides a schematic of the demand and supply for plantation forests. As the diagram shows, if the costs of plantation establishment decrease from $Cost_0$ to $Cost_1$, the supply curve shifts downward from $S$ to $S'$, other things being constant, and the quantity of plantations increases from $Q_0$ to $Q_1$. The economic benefits are the cost savings, represented by the area between the two cost curves and bounded by the demand curve ($D$) on the right and the vertical axes on the left.

The clonal propagation methods that are prerequisite for use of this kind of gene have been developed for hardwoods but not yet for softwoods (conifers). Forest plantation owners incur substantial costs in the early period of a rotation in order to generate larger but discounted benefits at some future time. High-yield plantations have rotations of 6 to 30 years. To the extent that the costs of establishment can be reduced, net benefits can be achieved. Experts estimate that herbicide resistance would reduce the costs of plantation establishment by an average of about \$35 an acre (\$87/ha) for fast-growing softwoods (reduced costs of 15%) and an average of \$160 an acre (\$400/

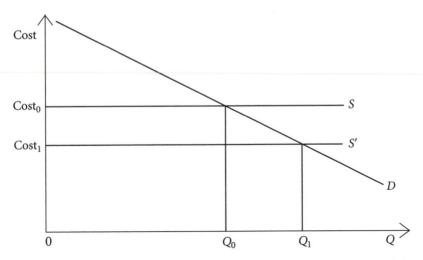

**FIGURE 3-2.** Schematic diagram of plantation establishment cost and supply ($S$) from plantation forests ($Q$) in relation to cost reductions that are expected to result from use of herbicide-resistant trees or other technologies.

ha) for fast-growing hardwoods (reduced costs of 30%) by eliminating the costs of other weed control activities. (The percentages are based on an updating of plantation establishment costs as found in Sedjo 1983.) In North America about 4 million acres are planted annually. If 98% (3.9 million acres) are softwood and 2% (0.1 million acres) hardwood, the potential cost reduction would be $136.5 million for softwoods and $16 million for hardwoods, or a total savings of $152.5 million annually.

Worldwide, about 10 million acres of plantation forest are established per year, roughly half conifer and half hardwood. If the amount of plantings were to remain unchanged, the potential savings from the introduction of the herbicide-resistance gene would be $175 million for softwoods and $800 million for hardwoods. Thus the potential global cost saving is roughly $975 million annually, assuming that low-cost conifer cloning is perfected. But even if softwoods are not included, the near-term potential benefits from planting trees with the herbicide-resistance gene are quite large.

Lower establishment costs are also expected to increase total plantation establishment. With 10 million hectares of forest planted annually, assuming that about 1 million hectares is new industrial plantations and that the actual costs to industry are reduced by the full amount of the cost reduction realized through the innovation (i.e., that the innovation is priced at marginal cost), that would mean an average reduction of 22.5% in plantation establishment costs.[1] How fast the annual rate of plantation establishment would rise would then depend in part on how responsive demand is to price

changes. That responsiveness is captured in the economist's use of price elas-
ticities (percentage change in quantity divided by the percentage change in
price). To examine this effect, I developed and estimated the impacts in the
following three scenarios:

**Scenario A: Maximum Impact.**   Given an initial total annual rate of global
planting of 1 million hectares, and assuming an infinite supply elasticity and
a unitary demand elasticity for forest plantation plantings (a derived
demand),[2] an estimated additional 225,000 hectares would be planted each
year. This assumes that the additional planting would reflect the current mix
of species and be divided evenly between conifers and hardwoods. Furthermore, if we assume that growth rates on plantation forests average 20 m$^3$ per
hectare per year for softwoods and 30 m$^3$ per hectare per year for hardwoods,
the additional plantings would add 2.5 million m$^3$ of wood volume per year
at harvest. If these increases in plantings were realized each year for 20 years,
about 100 million m$^3$ per year of additional industrial wood production
would be generated annually at the end of that period. At a 0.5% rate of
annual increase in consumption, on a 1997 production-consumption base of
1.5 billion m$^3$, global industrial wood consumption would be expected to
increase about 7.5 million m$^3$ annually.

**Scenario B: Intermediate Impact.**   Suppose the same conditions obtained as
in Scenario A, except that supply elasticity was 1.0. In this case a total of
112,500 additional hectares planted per year would increase the total production at harvest by 2.5 million m$^3$ per year. After 20 years of planting, this
would generate about 50 million m$^3$ of additional production each year.

**Scenario C: Minimum Impact.**   The assumption here is that supply elasticity
remains at +1.0, as in Scenario B, but that demand elasticity is –0.7 (approxi-
mating the recent FAO estimate of –0.67 for the elasticity of demand for
industrial roundwood. In this case an estimated 78,750 additional hectares
per year would be planted, with an increase in total production at harvest of
1.97 million m$^3$ per year. After 20 years of planting at this rate, the additional
wood production would be about 39 million m$^3$ per year.

## Environmental Benefits of Forest Biotechnology

Benefits can also be realized outside the market. Biotechnology in forestry can
be used to achieve some environmental goals and to mitigate various environmental problems. An important nonmarket benefit has already been realized
in the substitution of plantation-grown wood for the wood of primary forests.
Plantation wood can reduce the commercial logging pressure on natural for-

ests and thus on biodiversity "hot spots" and the natural habitats of important species (Sedjo and Botkin 1997). In fact, the rapid increase in the use of plantation wood, together with the stagnation of overall industrial wood production, suggests that wood from natural forests is declining both as a share of total industrial wood production and absolutely. Worldwide industrial wood production has been stable over the past two decades while total harvests from plantation forests have been increasing (Sohngen et al. 1999).

Modified tree species also show promise in providing environmental services in arid or drought-prone areas, saline soils, or frost zones where trees now may have difficulty surviving. Biotechnological modifications could allow trees to restore degraded lands, as well as provide traditional ecosystem services, such as erosion control and watershed protection, while increasing wood output. Transgenic trees could even be used for toxic bioremediation: Hybrid poplar, for example, has been genetically engineered to detoxify certain chemicals (see Bryson et al., this volume). Biotechnology also holds the potential to restore species severely damaged through pests and disease, such as the American chestnut, which was blighted around the turn of the twentieth century by an introduced fungus. Because the fungus acts only on the aboveground portions of the tree, live roots remain and could provide the bases for a restoration if the fungus were to be controlled through genetic engineering (Bailey 1997). Appropriate genes appear to be available in the Chinese chestnut. Finally, biotechnology could enhance the carbon sequestration ability of forests, and carbon-sequestering plantations in regions not currently forested could become important biological "sinks" that would mitigate the buildup of greenhouse gases associated with global warming (Kauppi et al. 2001).

## Conclusions

The distinguishing feature of the introduction of technology is higher productivity—that is, an increase in output per unit of input. Such increase in economic efficiency implies that society gets more output for its expenditure of inputs. The above analysis suggests that the economic benefit associated with the introduction of only one transgene, the herbicide resistance gene, could be as much as $1 billion annually. This benefit, which arises from the reduced cost of establishment, would increase the rate of tree plantation establishment into the indefinite future and thus increase the supply of industrial wood at lower prices. Substantial economic benefits could likewise be derived from other biotechnological innovations, including transgenic trees with other traits.

The increased biological and economic productivity of planted forests has important positive spillovers to the environment. As planted forest productivity and the number of low-cost plantation forests increase, the lower-cost industrial wood that those plantations produce will develop a greater comparative cost advantage over wood harvested from natural forests. Thus, as harvests from planted forests increase, production from natural forests declines; plantation wood is substituted for natural forest wood; and the natural forests are left for other uses, including ecosystem and biodiversity preservation.[3] This environmental benefit would occur through changes in the competitive structure of the forest industry. As the competitive advantage of natural and old-growth timber declines, the increased substitution of plantation wood gives natural forests a degree of protection from commercial logging because they are viewed as having more environmental value.

Finally, the biotechnological modification of a tree may allow it to perform a broader and more useful set of environmental functions and services. These could include enhanced carbon sequestration and wood production in regions that currently cannot support forestry. Biotechnology may also achieve other desired environmental objectives, such as toxic cleanups, land restoration, watershed enhancement, and erosion control in degraded, frigid, or arid areas.

There has been considerable discussion of whether biotechnology applied to agriculture will increase the demand for land, thereby putting more pressure on natural habitats. Some recent work suggests that that is likely to be the case if the demand for agricultural products is elastic (e.g., Angelsen and Kaimowitz 1998).[4] However, that is unlikely to be a problem in forestry, where demand is almost always estimated to be inelastic and the productivity of planted forests is considerably greater than that of natural forests.

## Notes

1. Sedjo (1999) estimated this to be about 600,000 ha for the tropics and subtropics; the model of Sohngen et al. (1999) estimated new plantations to be about 850,000 ha annually. The higher figure used in the latter study reflects the inclusion of new plantation establishment in the temperate regions and anecdotal evidence suggesting that earlier estimates were on the modest side.

2. Derived demand refers to the demand for an intermediate good, used in the production of a consumption good, where the demand for the intermediate good is "derived" from the final demand for the consumption good and the characteristics of production.

3. The argument that plantation wood substitutes for wood from natural forests is substantially different from the questions surrounding land involved in grain production, because forestry has been a foraging activity, not a cropping activity. In a study for the FAO, Brooks et al. (1996) estimated the elasticity of global demand for industrial wood at –0.67.

4. It has been noted that since cattle are increasingly fattened in feedlots, where they consume grains, the total demand for grain, human and animal, may be elastic.

This implies that if grain prices fall—say, because of biotechnology—the total area of land in grains could increase. However, it should be noted that where both grain and beef are part of the diet, the feeding of grain to cattle has resulted in a decline in pasture area. Thus, total agricultural land, grain plus pasture, may have decreased even if the area in grains has increased.

# References

Angelsen, A., and D. Kaimowitz. 1998. When Does Technological Change in Agriculture Promote Deforestation? Paper presented at the AAEA International Conference on Agricultural Intensification, Economic Development and Environment. Salt Lake City, July 31–August 5.

Bailey, R. 1997. American Chestnut Foundation. Washington, DC: Center for Private Conservation, Competitive Enterprise Institute. http://www.cei.org/gencon/025,01358.cfm (accessed June 2004).

Bradshaw, T. 1999. A Blueprint for Forest Tree Domestication. Seattle: College of Forest Resources, University of Washington. http://poplar2.cfr.washington.edu/toby/treedom.htm (accessed July 2002).

Brooks, D., H. Pajulja, T. Peck, B. Solberg, and P. Wardle. 1996. Long-Term Trends and Prospects in the Supply and Demand for Timber and Implications for Sustainable Forest. Rome, Italy: FAO.

FAO (UN Food and Agricultural Organization) N.D. *Forest Products Yearbook,* selected issues. Rome, Italy: FAO.

Kauppi, P., et al. 2001. Technical and Economic Potential of Options to Enhance, Maintain and Manage Biological Carbon Reservoirs and Geo-engineering. In *Climate Change 2001: Mitigation,* edited by B. Metz, et al. Contribution of Working Group III, Third Assessment Report on Climate Change. Cambridge, UK: Intergovernmental Panel on Climate Change (IPCC).

McLean, M.A., and P.J. Charest. 2000. The Regulation of Transgenic Trees in North America. *Silvae Genetica* 49(6): 233–239.

Mullin, T.J., and S. Bertrand. 1998. Environmental Release of Transgenic Trees in Canada—Potential Benefits and Assessment of Biosafety. *Forestry Chronicle* 74(2): 203–219.

Sedjo, R.A. 1983. *The Comparative Economics of Plantation Forests.* Washington, DC: Resources for the Future.

————. 1999. Biotechnology and Planted Forests: Assessment of Potential and Possibilities. RFF Discussion Paper 00-07. Washington, DC: Resources for the Future.

Sedjo, R.A., and D. Botkin. 1997. Using Forest Plantations to Spare Natural Forests. *Environment* 39(10): 15–20, 30.

Sedjo, R.A., A. Goetzl, and M. Steverson. 1997. *Sustainability of Temperate Forests.* Washington, DC: Resources for the Future.

Sohngen, B., R. Mendelson, and R.A. Sedjo. 1999. Forest Management, Conservation, and Global Timber Markets. *American Journal of Agricultural Economics* 81(1): 1–13.

Yanchuk, A.D. 2001. The Role and Implication of Biotechnological Tools in Forestry. *Unasylva* 204(52): 53–61.

# 4

# Accomplishments and Challenges in Genetic Engineering of Forest Trees

RICK MEILAN, DAVE ELLIS, GILLES PILATE,
AMY M. BRUNNER, AND JEFF SKINNER

In many respects, the process of stably integrating a piece of DNA (i.e., a transgene) into a tree's nuclear genome is the same as with traditional agricultural crops. There are numerous examples of successful transformation in both. The first commercial release of a transgenic tree occurred several years ago in Hawaii with papaya (Gonsalves 1998). It not only demonstrates the efficacy of this approach for the genetic improvement of trees but also serves to answer many of the questions regarding the stability of transgene expression in trees. As has occurred with other agricultural crops, papaya growers have enthusiastically embraced biotechnology, as evidenced by the rapid adoption of the product and continued satisfaction with it. Although numerous transgenic crops have been commercialized (see http://www.aphis.usda.gov/biotech/), papaya has been the only transgenic perennial released.

At present there is a shortage neither of genes nor of methods for incorporating them into trees, and the genes that are inserted are expressed adequately. What we currently lack is the ability to coax individual cells from commercially important genotypes to grow into whole plants. This is not an issue with transgenics per se but is instead a tissue culture problem. It is not a problem that is unique to trees but exists with some traditional agricultural crops as well.

Why then are there far more commercial releases of genetically modified agricultural crops than of trees? Economics is part of the answer, but fundamental biology also plays a role. In annual agricultural crops, tissue culture systems need only be developed for a small number of breeding lines, and then the transgene can be rapidly incorporated into elite lines through traditional breeding. For example, in maize, even though hundreds of different transgenic varieties are available and transgenics accounted for 26% of the acreage planted to maize in 2001, all of these varieties were derived from only five transgenic events in as few as two breeding lines.

The ability to introgress a transgene through breeding is key to the very rapid development of engineered crop varieties. In maize, for example, a single transformation event can theoretically be introgressed from a breeding line into an elite line in nine generations, or as little as three years. This enables all of the various agronomic crop cultivars, all specifically adapted to different locations or environments, to have the same reliable expression of a trait imparted by a single insertion event. In addition, the resulting varieties can be readily amplified for rapid deployment. Equally important is that because of the ease with which a transgene can be introgressed, tissue culture and transformation systems are not needed for each commercial crop variety.

This is in stark contrast to what occurs with trees, where introgressing a transgene into superior genotypes through breeding is not currently possible. Even if the long regeneration cycle in trees could be overcome, the inbred lines or breeding lines required for such an introgression program have not been developed. At present the only option with respect to trees is to develop robust transformation systems that are adaptable to the many genotypes used in clonal plantation forestry.

However, tissue culture systems for most of the genotypes that are commercially important in forestry are not available. Each genotype requires a slightly different set of conditions to induce individual cells containing the inserted DNA to differentiate into whole plants. Currently our ability to identify the specialized conditions needed for individual genotypes is limited.

Generalities in the literature have led researchers to assume that numerous tree species can be transformed. It should be emphasized, however, that in many cases only a *single* genotype of a species has been transformed. Poplar, which has been touted as the "model tree" for molecular studies, provides a good example. It has a small genome, and by all indications it can be transformed at a high frequency. Yet a careful examination of the reports on field trials of transgenic poplars reveals that a majority of lines described have been produced in just a handful of genotypes. Furthermore, only a few, if any, of those genotypes are grown commercially. Thus, the difficulty in manipulating commercially important genotypes in tissue culture remains an obstacle to gaining the full economic benefit from transgenics in all forestry species.

Despite this limitation, there are over 100 reports of trees' being transformed with valuable traits, including herbicide and insect resistance (Brasileiro et al. 1992; Ellis et al. 1993; Donahue et al. 1994; Cornu et al. 1996; Meilan et al. 2000; 2002); modification of lignin (Hu et al. 1999; Pilate et al. 2002); glutathione metabolism (Foyer et al. 1995; Strohm et al. 1995); modification of cellulose (Shani et al. 2001); bioremediation (Rugh et al.

1998; Shang et al. 2001); and altered hormone biosynthesis (Eriksson et al. 2000). In this chapter we discuss some of the challenges, both perceived and real, to the use of transgenes in long-lived perennial crops.

## Gene Expression

A question frequently asked about long-lived perennial crops such as trees is, How can one be sure that transgenes will continue to be expressed over the years or decades required to complete a rotation? To answer this question, one must look at the data available from transgenic crops, both annual and perennial.

In any discussion on the expression of transgenes, it is important to realize that variation and instability in transgene expression do exist, just as variation in gene expression exists in conventional breeding. Indeed, if variation in gene expression did not exist, breeding programs would be severely hampered. Moreover, just as in breeding, if one looks hard enough it is possible to find virtually any pattern of gene expression, from suppression to high-level expression. Fortunately, in both transgenics and breeding, stable gene expression is very reliable. Were that not true, there would be no commercial transgenic crops, and the industry would not have advanced to its present state.

Historically, primary selections have been done based on the desired level or location of transgene expression. Recently there has been a shift toward selection for simpler insertion patterns, and in many agricultural crops single insertion events are now the first criterion used to select transformed lines for further study. This change has been made mainly to satisfy regulatory requirements for complete molecular characterization, including sequence information, for each insert and its flanking genomic DNA (i.e., characterizing a single site requires less effort than multiple sites). It also greatly facilitates introgression of transgenes into diverse crop varieties.

We know of no reproducible evidence to support the widespread belief that there is increased variation in expression patterns with multiple insertion sites. Discussions with colleagues working in agricultural biotechnology have repeatedly confirmed that there is little support for that notion. Rather, the available data suggest that the desired level of transgene expression occurs frequently enough to obviate the need to use insert complexity as the primary criterion for selecting transgenic events, at least in asexually propagated crops.

The foregoing discussion addresses variation in transgene expression, but more important is the question of the stability of transgene expression over time in long-lived crops. As with expression level, it is easy to speculate that biotechnologists select for simple patterns in hopes of achieving more stable expression. We find no reliable documentation to support the belief that simple transgene insertion patterns lead to more stable expression over time.

It is of paramount importance that transgenes are expressed in the expected way throughout the life of a tree. In agricultural crops, there is no question that such stability exists; markets would not have expanded if transgene expression levels could not be trusted.

Although no studies have yet been conducted over a full rotation for any forest tree, there is abundant evidence that transgene expression is very stable in trees over time. In one of the earliest studies of transgenic trees, Ellis and colleagues (unpublished) measured the expression of *GUS* (a visual reporter gene) in field-grown transgenic poplar and spruce over a three-year period. They observed that variation in expression levels among individuals within a single line were greater in the field than in tissue culture. They speculated that similar trends in gene expression would also have been noted had a native gene been studied and that environmental factors affecting gene expression were minor in tissue culture and far greater in a field setting. However, despite the observed within-line variation, differences *between* lines were far more significant, and the between-line differences were consistent throughout environments and between years. Of greater importance was the observation that during the three-year study, the level of *GUS* expression was not significantly different ($p < .05$) from year to year for approximately 85% of the 15 annual sample dates.

In another aspect of this same study, transgene expression was measured over several years in poplars that contained *GUS* under the control of the promoter from a wound-inducible gene, *PINII*. In these trees, wound-inducible *GUS* expression was found to be consistent over the two-year field study, and the level of *GUS* expression was faithfully induced in a very strict developmental pattern throughout both years. These data demonstrate that transgenes need not be expressed continuously to be reliably expressed in trees at specific developmental stages.

Meilan et al. (2001b) assessed the stability of transgene expression in 40 transgenic lines (i.e., independent events) of hybrid cottonwood (*Populus trichocarpa* x *P. deltoides*) grown at three field sites during four years of field trials. All lines were transformed with a binary vector that included two genes conferring tolerance to glyphosate (*GOX* and *CP4*), a gene encoding resistance to the antibiotic kanamycin (*NPTII*), and *GUS*. *Agrobacterium tumefaciens* was used for transformation; callogenesis and organogenesis occurred under kanamycin selection. To test the stability of transgene expression, they repeatedly applied herbicide to all lines after outplanting, challenging ramets from previously untreated lines during their fourth season of vegetative growth. They used maintenance of herbicide tolerance and *GUS* expression as indicators of transgene stability. Their data show that all lines that were highly tolerant in year one continued to be highly tolerant in year four.

In what is perhaps the most ambitious study done to date with transgenic trees, Hawkins et al. (2002) examined transgene expression in field-grown transgenic poplar for a decade. They also observed very stable and dependable *GUS* expression from year to year. More important, they found no evidence of transgene rearrangements, multiplication, loss, or other modifications. These data are the strongest indication thus far that transgene expression is stable in long-lived perennial crops such as forest trees.

Han et al. (1997) evaluated the use of matrix-attachment region (MAR) elements derived from a tobacco gene for increasing the frequency of *Agrobacterium*-mediated transformation. MARs are elements of DNA that can enhance and stabilize transgene expression (Spiker and Thompson 1996). A binary vector that carried the *GUS* reporter gene containing an intron and an *NPTII* gene was modified to contain flanking MAR elements within the T-DNA borders. The MAR-containing vectors were used to transform tobacco, a readily transformable poplar clone (*P. tremula* x *P. alba*), and a recalcitrant poplar clone (*P. trichocarpa* x *P. deltoides*). MARs significantly enhanced transgene expression and transformation efficiency, but the effects varied widely in magnitude among genotypes. MARs increased *GUS* gene expression approximately tenfold in the two hybrid poplar clones and doubled it in tobacco one month after cocultivation with *Agrobacterium*; they increased the frequency of kanamycin-resistant poplar shoot recovery more than eightfold. In a more recent report, it was shown that MARs did not reduce variability in transgene expression, but transgene silencing was reduced and transgene expression was elevated (Brouwer et al. 2002). Thus, MARs hold considerable promise for use with poplar and presumably other trees as well.

Despite these encouraging results, it has been suggested that the best evidence for expression stability is maintenance through meiosis. Although use of primary transformants in proven elite genotypes is the current focus of transgenic research, future tree improvement programs could benefit from the inclusion of transgenes in sexual breeding populations, and therefore, the question of transgene expression in progeny is of practical significance. In the only known example of transgenic inheritance in forest trees, Gilles Pilate and colleagues (INRA, Orléans, France, unpublished data) observed the expected Mendelian segregation ratio for progeny from a line containing a single-copy insert. Moreover, these workers detected a fragment of the expected size hybridizing to a *GUS* probe in all kanamycin-resistant lines. In contrast, a line containing four inserts produced progeny containing one, two, three, and four copies of the segregated transgene. Transgene expression level was highly variable among the progeny from both transformed lines, but that is to be expected from heterozygous offspring.

In addition to variation in the level or timing of transgene expression, there have also been numerous reports of transgene silencing, and the phenomenon has been reviewed extensively (e.g., Matzke et al. 2002). One form of silencing occurs at the level of transcription and usually involves methylation of promoter regions, although it may also involve chromatin modification (van

Blokland et al. 1997). These reports have led many to conclude erroneously that silencing is a major problem confronting genetic engineering. As mentioned in the beginning of this section, it is possible to find all expression patterns, and gene silencing is no exception. We believe that there are many reports of transgene silencing in the literature because it is scientifically interesting. However, it is a rare occurrence in our hands and is not a problem with the promoters, genes, or constructs currently being tested in trees.

Hawkins et al. (2002) observed only one case of transgene silencing in their work, an event that occurred after selection but prior to any analysis of the transformed lines. Interestingly, that case is similar to the only case of transgene silencing that Ellis et al. (1993) saw in poplar, where again the putative silencing event was found early in tissue culture. In both cases, 5-azacytidine had no effect on the silencing, indicating that either methylation was not involved, or the silencing was not due to an azacytidine-reversible methylation event. These two incidents of gene silencing, while of academic interest, occurred in less than 1% of all transgenic lines analyzed.

Meilan et al. (2002) conducted a two-year field test in which they evaluated 80 transgenic lines for genetically engineered herbicide tolerance. They observed an abrupt increase in mean herbicide damage for two of the lines from one year to the next. Quantitative transgene silencing is one possible explanation for loss of tolerance. A recent report described how cold-induced dormancy led to elevated methylation, which in turn led to partial transgene silencing (Callahan et al. 2000). The possibility of year-to-year variation in transgene expression highlights the importance of conducting multiyear trials. All 80 lines were vegetatively propagated and have been planted on another site to monitor their growth and herbicide tolerance for several years.

Contrary to what we have collectively seen in our work, there is recent evidence for a high level of co- or sense-suppression in poplar with constructs used to overexpress genes involved in lignin biosynthesis. Examples of genes that have been cosuppressed include *COMT, CAD, 4CL,* and *C4H* (Jouanin et al. 2001; Tsai et al. 2001). It is not known whether this suppression is the result of using homologous genes and will thus become more frequent as more tree genes are used. It is feasible that the silencing phenomenon could be unique to the lignin biosynthesis. It will be interesting to see if silencing is a function of the biochemical pathway being manipulated or the source of the genes used to manipulate the pathway, or both.

## Somaclonal Variation

Much like transgene silencing, somaclonal variation, a stable genetic change in somatic cells of a plant, does occur but can be minimized or avoided with careful control over cultural conditions. Somaclonal variation tends to be more common with unstable genomes such as polyploids. But these are very rare among forest trees.

In one study on the induction of somaclonal variation in poplar, Ostry et al. (1994) started with a single mother plant and compared regenerated shoots derived from embryogenic cell suspensions, protoplasts, leaf micro-cross-sections, callus, and shoot or root cultures. Somaclonal variation was scored using morphological characters and increased resistance to *Septoria*, a fungal pathogen, based on a leaf disc assay. Morphological variants were noted in adventitious shoots regenerated from roots, callus, and protoplasts. In this study, shoots from stem callus yielded the highest level of *Septoria*-resistant plants, indicating that that treatment yielded the highest level of putative somaclonal variants. In a follow-up report, Ostry and Ward (2003) described their somaclones as stable through vegetative propagation and in field plots for up to 11 years. These data confirm earlier work suggesting that the longer cells are maintained in an unorganized culture, the greater the likelihood of somaclonal variation occurring (Deverno 1995).

In a related study in poplars (Serres and McCown, University of Wisconsin-Madison, unpublished), plants from approximately 400 colonies derived from individual protoplasts were regenerated and grown ex vitro. After 10 weeks in a greenhouse, the plants were scored for 17 different parameters ranging from leaf size and shape to internode length. In all, nearly 50% of the regenerated plants had some morphological differences from the control, with the greatest number of variants being in height, leaf length/width ratio, stem diameter, and number of nodes. These changes were stable in the green-house, and most remained throughout the first season when the plants were grown in the field. The plants were allowed to overwinter outdoors and were scored again the following spring after leaf flush. In the second year, only three lines continued to display abnormalities; two lines showed sectoring in the leaves, perhaps caused by a transposition event, and the remaining line had the appearance of a tetraploid (Serres et al. 1991). Clearly the majority of the variations were transient and not true somaclonal changes.

Meilan et al. (unpublished) have produced more than 5,000 independent transgenic lines in 17 different poplar genotypes (14 of cottonwood, sections Tacamahaca and Aigeiros, and 3 of aspen, section Populus or Leuce). They field-tested 557 of those lines and have grown most of the remainder in the greenhouse for extended periods. Of the total, they have observed obvious morphological abnormalities that were not induced by transgene expression in only three lines (0.06%), none in hybrid aspen. The three lines in which they observed putative somaclonal variants were all hybrid cottonwoods, whose transformation protocol requires that they spend nearly twice as long in an undifferentiated state as do the aspens (Han et al. 2000).

These studies are consistent with the view that (a) minimizing the duration of an unorganized callus stage and (b) keeping the use of hormones to maintain or regenerate shoots as brief as possible are important precautions for reducing somaclonal variation. It should be noted, however, that even in the case of protoplasts, where thiadiazuron and both an auxin and a cytokinin were used for regeneration, less than 2% of the regenerated shoots exhibited stable and confirmed somaclonal variation. It appears that with careful attention to culture conditions and selection of healthy transgenic lines, somaclonal variation is not an important problem for most kinds of transgenic plants.

## Transgene Containment

The ecological need for strict transgene containment within plantations, generally assumed to be achieved via engineered sterility, will depend on the trait, the environment within which the transgenics will be grown, and the species. In addition, social, political, and ethical considerations may dictate whether a high level of containment is sought. Each case must be considered individually.

With the tools that are currently available, absolute sterility in plants is unlikely to be technically achievable in the foreseeable future. Thus it is important to assess whether sterility is truly needed and what level will suffice for specific applications. There will also be many cases in which it would be desirable to incorporate transgenes into a conventional breeding system or to preserve flowers and fruits intact for the benefit of wildlife. Nonetheless, sterility will be an important genetic engineering tool for many transgenic applications.

Genetically engineering reproductive sterility in trees is desirable for several reasons (Meilan et al. 2001a). First, the development of trees that are incapable of producing sexual propagules would limit gene flow into the wild, helping to mitigate ecological concerns over establishment of transgenic plantations. Second, it will likely lessen the slowing of growth associated with the onset of maturation (Eis et al. 1965; Tappeiner 1969). Third, it could eliminate the production of pollen and other nuisance reproductive structures.

One common way to engineer sterility is to ablate cells by expressing a deleterious gene in a tissue-specific fashion. Floral tissue-specific promoters are fused to one of a variety of cytotoxin genes that lead to rapid and early death of the cells within which the gene product is expressed. One of the more popular ways to engineer sterility in herbaceous plants employs an RNAse gene that, although isolated from a bacterium, encodes a type of enzyme that is common in plants and animals (Mariani et al. 1990).

A second way to genetically engineer flowering control is through the use of dominant negative mutations (DNMs). DNM genes encode mutant proteins that suppress the activity of coexisting wild-type proteins (Espeseth et al. 1993). Inhibition can occur by a variety of means, including formation of an inactive heterodimer, sequestration of protein cofactors, sequestration of metabolites, or stable binding to a DNA regulatory motif. The potential of this approach for floral control was demonstrated by studies of a rice MADS-box gene (Jeon et al. 2000). Overexpression of genes that encode proteins with amino acid changes in the highly conserved MADS domain resulted in mutant floral phenotypes. Similar changes, which eliminated the encoded protein's ability to bind DNA, were previously reported to produce DNMs in mammalian MADS-box genes (Molkentin et al. 1996). A potentially power-ful alternative approach is to introduce a transgene encoding a zinc-finger protein specifically designed to block transcription of the target gene (Beerli and Barbas 2002).

A third technique to control flowering involves gene silencing. Recent studies in a variety of eukaryotic organisms have shown that double-stranded RNA (dsRNA) is an inducer of homology-dependent gene silencing, and use of dsRNA to induce silencing has been termed "RNA interference" (RNAi) (Hannon 2002). Studies in plants have shown that strong silencing can be achieved through the introduction of a transgene containing an inverted repeat of a sequence corresponding to part of the transcribed region of the endogenous gene targeted for silencing (e.g., Chuang and Meyerowitz 2000). Such transgenes induce posttranscriptional gene silencing (PTGS) by trigger-ing RNA degradation. Although this approach appears to provide a reliable means for engineering stable suppression of gene activity in plants, whether PTGS will be effective practically is uncertain, due in part to the ability of plant viruses to suppress PTGS (reviewed in Voinnet 2001). An alternative is to use a transgene containing an inverted repeat of a target gene's promoter region, which has been shown to induce de novo DNA methylation and tran-scriptional gene silencing (TGS) (Aufsatz et al. 2002). Unlike PTGS, TGS is not susceptible to viral suppression; however, it is unclear whether all endog-enous plant promoters can be silenced by this method.

In early attempts at using cell ablation with poplar, heterologous promot-ers, which had shown floral-specific expression in tobacco and *Arabidopsis* (Koltunow et al. 1990; Hackett et al. 1992), were used to drive the expression of two cytotoxin genes, *DTA* (Greenfield et al. 1983) and barnase (Hartley 1988). When introduced into transgenic poplars, these fusions resulted in decreased vegetative growth, suggesting leaky expression in nontarget tissues (Meilan et al. 2001a).

Now that floral homeotic genes from poplar have been cloned and charac-terized (Brunner et al. 2000; Sheppard et al. 2000; Rottmann et al. 2000), work has begun on experimentation with these promoters from poplar genes. The promoter from PTD (the *P. trichocarpa* homolog of DEFICIENS) appears to be the most floral-specific in its expression pattern (Sheppard et

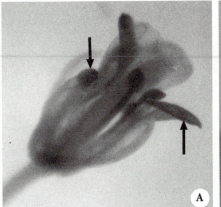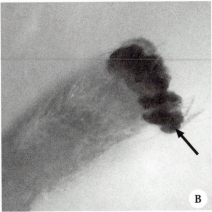

**FIGURE 4-1.** PTD-directed *GUS* expression patterns in *Arabidopsis* (A) and poplar (B) flowers. The staminate poplar flower is from a hybrid aspen clone (INRA 353-38) cotransformed with 35S-LFY to induce early flowering. Darkened regions (arrows) reveal tissues within which *GUS* was expressed. In *Arabidopsis*, expression was confined to petals and stamens; in poplar, expression was primarily in the stamens.

al. 2000). We have shown that the PTD promoter directs expression of the *GUS* gene early in the development of floral organs in Arabidopsis and poplar (see Figure 4-1). The latter was cotransformed with the LEAFY (LFY) gene from Arabidopsis under the control of the 35S promoter, which has been shown to induce early flowering in poplar (Weigel and Nilsson 1995). When the PTD promoter was used to drive the expression of a cytotoxin gene, petals and stamens were ablated in Arabidopsis; petals, stamens, and carpels were absent in transgenic tobacco (Figure 4-2). PTD::*DTA* also prevented flowers from forming on poplar cotransformed with 35S::LFY. Expression of PTD::*DTA* had no significant effects on growth in tobacco. These results suggest that the PTD promoter may be useful for engineering sterility in a variety of species. Studies are also under way to use RNAi and DNM approaches to suppress several floral homeotic genes in transgenic poplars (Brunner and Meilan, unpublished data).

## Conclusion

In summary, numerous field tests conducted worldwide clearly demonstrate that it is possible to genetically engineer diverse tree species. Despite these results, the main challenge with trees continues to be the development of efficient transformation systems for the most desired genotypes. In this regard, it is not transformation per se that is limiting, but the tissue culture

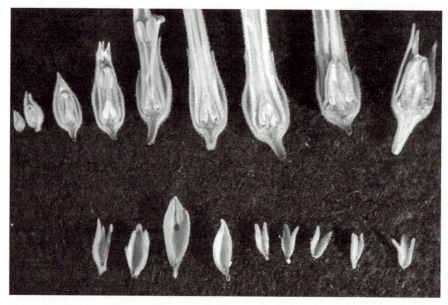

**FIGURE 4-2.** PTD-directed *DTA* expression in tobacco. Longitudinal cross-sections showing developmental sequence of flowers from nontransgenic plants (upper tier) and plants transformed with PTD::*DTA* (lower tier). Flowers from the PTD::*DTA* plants had all generative organs ablated and consisted solely of sepals.

system needed to regenerate whole trees from single cells containing the inserted genes. Although more research is needed, results from field tests have been most valuable for improving our understanding of how inserted genes will be expressed in long-lived perennials.

Available evidence suggests that the issues associated with transgene expression in trees will be the same as those observed in agronomic crops. To date we have seen that (a) transgene expression levels vary among transformation events/lines, species, transgenes, and traits being modified; (b) targeted, tissue-specific expression of transgenes is possible; (c) transformation events that give consistent levels of gene expression in primary transformants are the rule, not the exception; and (d) it is possible to find whatever is being sought, whether it be gene silencing or commercially useful expression. These trends suggest that although additional work is needed in the area of temporal and developmental regulation, stable, long-term expression of transgenes in trees is readily achievable.

Similarly, somaclonal variation does not appear to be a widespread phenomenon when regenerating transgenic trees. The use of embryogenesis helps reduce its incidence with gymnosperms, and with available practices and care in tissue culture, somaclonal variation is not a limitation with angiosperms.

Another factor that will play a role in the use of genetically modified trees is the possibility of undesirable levels of transgene spread. Environmental risks associated with transgene movement are specific to the gene(s) being inserted, the tree species, and the environment within which the transgenic tree is planted. Clearly, pollen and seed have the greatest potential to facilitate transgene movement, but whether they do so depends on the reproductive biology of the species in question, the management scheme being used, and the proximity of transgenic trees to sexually compatible relatives. Although cultural and deployment strategies could be used in some locations to minimize the risk of transgene spread, another approach is to link genes that inhibit floral development with the transgene of interest. Several labs are using this approach to develop methods for transgene containment, and it should soon be possible to genetically engineer flowering control in trees.

Thus it has been shown that we can produce transgenic trees with almost no evidence of collateral genetic damage and that inserted genes are expressed stably from year to year, after vegetative propagation, and in a variety of environments. However, these results were collected from transgenic trees grown in numerous small-scale, short-term field trials. We believe it is now time to conduct long-term ecological studies on a larger scale to evaluate potential risks of transgenics to the environment.

## Acknowledgments

The work described herein was funded in part by the Tree Genetic Engineering Research Cooperative (http://www.fsl.orst.edu/tgerc/index.htm); the U.S. Department of Energy's Biomass Program, through contract 85X-ST807V with Oak Ridge National Laboratory, which is managed by UT-Battelle, LLC, for the U.S. Department of Energy under contract DE-AC05-00OR22725; National Science Foundation I/UCRC Program (grant #9980423-EEC); Consortium for Plant Biotechnology Research (grant #OR22072-78); Agenda2020 (grant #FC0797ID13552); a grant from the Monsanto Company; and by EU project AIR2-CT94-1571. The authors would also like to express their gratitude to Caiping Ma, Vicky Hollenbeck, Sarah Dye, Céline Pugieux, and Simon Hawkins for their help with the work; to Wayne Parrot, Maureen Fitch, and Janet Carpenter for valuable information and discussion; and to Margarita Gilbert for comments on the manuscript.

## References

Aufsatz, W., M.F. Mette, J. Van Der Winden, M. Matzke, and A.J. Matzke. 2002. HDA6, A Putative Histone Deacetylase Needed to Enhance DNA Methylation Induced by Double-Stranded RNA. *EMBO Journal* 21(24): 6832–6841.

Beerli, R.R., and C.F. Barbas III. 2002. Engineering Polydactyl Zinc-Finger Transcription Factors. *Nature Biotechnolgy* 20: 135–141.

Brasileiro, A.C.M., C. Tourneur, J.-C. Leplé, V. Combes, L. Jouanin. 1992. Expression of the Mutant *Arabidopsis thaliana* Acetolactate Synthase Gene Confers Chlorsulfuron Resistance to Transgenic Poplar Plants. *Transgenic Research* 1: 133–41.

Brouwer, C., W. Bruce, S. Maddock, Z. Avramova, and B. Bowen. 2002. Suppression of Transgene Silencing by Matrix Attachment Regions in Maize: A Dual Role for the Maize 5' ADH1 Matrix Attachment Region. *Plant Cell* 14(9): 2251–2264.

Brunner, A.M., W.H. Rottmann, L.A. Sheppard, K. Krutovskii, S.P. DiFazio, S. Leonardi, and S.H. Strauss. 2000. Structure and Expression of Duplicate *AGAMOUS* Orthologs in Poplar. *Plant Molecular Biology* 44: 619–634.

Callahan, A.M., R. Scorza, L. Levy, V.D. Damsteegt, and M. Ravelonandro. 2000. Cold-Induced Dormancy Affects Methylation and Post-transcriptional Gene Silencing in Transgenic Plums Containing Plum Pox Potyvirus Coat Protein Gene. Paper presented at International Society of Plant Molecular Biology Congress, June 18–24, Québec, Canada.

Chuang, C.-F., and E.M. Meyerowitz. 2000. Specific and Heritable Genetic Interference by Double-Stranded RNA in *Arabidopsis thaliana. Proceedings National Academy of Sciences USA* 97(9): 4985–4990.

Cornu, D., J.-C. Leplé, M. Bonnadé-Bottino, A. Ross, S. Augustin, A. Delplanque, L. Jouanin, and G. Pilate. 1996. Expression of a Proteinase Inhibitor and a *Bacillus thuringiensis* d-Endotoxin in Transgenic Poplars. In *Somatic Cell Genetics and Molecular Genetics of Trees*, edited by M.R. Ahuja, W. Boerjan, and D.B. Neal. Dordrecht, Netherlands: Kluwer, 131–136.

Deverno, L.L. 1995. An Evaluation of Somaclonal Variation during Somatic Embryognesis. In *Somatic Embryogenesis in Woody Plants*, vol. 1, edited by S. Jain, P. Bupta, and R. Newton. Dordrecht, Netherlands: Kluwer, 361–377.

Donahue, R.A., T.D. Davis, C.H. Michler, D.E. Riemenschneider, D.R. Carter, P.E. Marquardt, N. Sankhla, D. Sankhla, B.E. Haissig, and J.G. Isebrands. 1994. Growth, Photosynthesis, and Herbicide Tolerance of Genetically Modified Hybrid Poplar. *Canadian Journal of Forest Research* 24(12): 2377–2383.

Eis, S., E.H. Garman, and L.F. Ebell. 1965. Relation between Cone Production and Diameter Increment of Douglas-Fir [*Pseudotsuga menziesii* (Mirb.) Franco], Grand Fir [*Abies grandis* (Dougl.) Lindl.], and Western White Pine [*Pinus monticola* (Dougl.)]. *Canadian Journal of Botany* 43: 1553–1559.

Ellis, D., D. McCabe, S. McInnis, R. Ramachandran, D. Russell, K. Wallace, B. Martinell, D. Roberts, K. Raffa, and B. McCown. 1993. Stable Transformation of *Picea glauca* by Particle Acceleration. *Bio/Technology* 11(1): 84–89.

Eriksson, M.E., A. Israelsson, O. Olsson, and T. Moritz. 2000. Increased Gibberellin Biosynthesis in Transgenic Trees Promotes Growth, Biomass Production and Xylem Fiber Length. *Nature Biotechnology* 18: 784–788.

Espeseth, A.S., A.L. Darrow, and E. Linney. 1993. Signal Transduction Systems: Dominant Negative Strategies and Mechanisms. *Molecular and Cellular Differentiation* 1: 111–161.

Foyer, C.H., N. Souriau, S. Perret, M. Lelandais, K.J. Kunert, C. Pruvost, and L. Jouanin. 1995. Overexpression of Glutathione Reductase But Not Glutathione Synthetase Leads to Increases in Antioxidant Capacity and Resistance to Photoinhibition in Poplar Trees. *Plant Physiology* 109: 1047–1057.

Gonsalves, D. 1998. Control of Papaya Ringspot Virus in Papaya: A Case Study. *Annual Review of Phytopathology* 36: 415–437.

Greenfield, L., M.J. Bjorn, G. Horn, D. Fong, G.A. Buck, R.J. Collier, and D.A. Kaplan. 1983. Nucleotide Sequence of the Structural Gene for Diphtheria Toxin Carried by Corynebacteriophage. *Proceedings of the National Academy of Sciences USA* 80: 6853–6857.

Hackett, R.M., M.J. Lawrence, and C.H. Franklin. 1992. A *Brassica* S-Locus Related Gene Promoter Directs Expression in Both Pollen and Pistil of Tobacco. *Plant Journal* 2: 613–617.

Han, K.-H., C. Ma, and S.H. Strauss. 1997. Matrix Attachment Regions (MARs) Enhance Transformation Frequency and Transgene Expression in Poplar. *Transgenic Research* 6: 415–420.

Han, K.-H., R. Meilan, C. Ma, and S.H. Strauss. 2000. An *Agrobacterium* Transformation Protocol Effective in a Variety of Cottonwood Hybrids (Genus *Populus*). *Plant Cell Reports* 19:3 15–320.

Hannon, G.J. 2002. RNA Interference. *Nature* 418: 244–251.

Hartley, R.W. 1988. Barnase and Barstar: Expression of Its Cloned Inhibitor Permits Expression of a Cloned Ribonuclease. *Journal of Molecular Biology* 202: 913–915.

Hawkins S., J.C. Leplé, D. Cornu, L. Jouanin, and G. Pilate. 2002. Stability of Transgene Expression in Poplar: A Model Forest Tree Species. *Annals of Forest Science* (in press).

Hu, W.-J., S.A. Harding, J. Lung, J.L. Popko, J. Ralph, D.D. Stokke, C.J. Tsai, and V.L. Chiang. 1999. Repression of Lignin Biosynthesis Promotes Cellulose Accumulation and Growth in Transgenic Trees. *Nature Biotechnology* 17: 808–812.

Jeon, J.-S., S. Jang, S. Lee, J. Nam, C. Kim, S.-H. Lee, Y.-Y. Chung, S.-R. Kim, Y.H. Lee, Y.-G. Cho, and G. An. 2000. *Leafy hull sterile1* Is a Homeotic Mutation in a Rice MADS Box Gene Affecting Rice Flower Development. *Plant Cell* 12: 871–884.

Jouanin, L., T. Goujon, R. Sibout, B. Pollet, I. Mila, J.C. Leplé, G. Pilate, M. Petit-Conil, J. Ralph, and C. Lapierre. 2001. Lignification in Poplar and Arabidopsis Plants with Depressed COMT and CAD Activity. Paper presented at the Tree Biotechnology in the New Millennium IUFRO/Molecular Biology of Trees Meeting, July 22–27, Stevenson, WA.

Koltunow, A.M., J. Truettner, K.H. Cox, M. Wallroth, and R.B. Goldberg. 1990. Different Temporal and Spatial Gene Expression Patterns Occur during Anther Development. *Plant Cell* 2: 1201–1224.

Mariani, C., M. DeBeuckeleer, J. Truettner, J. Leemans, and R.B. Goldberg. 1990. Induction of Male Sterility in Plants by a Chimaeric Ribonuclease Gene. *Nature* 347: 737–741.

Matzke, M.A., W. Aufsatz, T. Kanno, M.F. Mette, and A.J. Matzke. 2002. Homology-Dependent Gene Silencing and Host Defense in Plants. *Advances in Genetics* 46: 235-275.

Meilan, R., C. Ma, S. Cheng, J.A. Eaton, L.K. Miller, R.P. Crockett, S.P. DiFazio, and S.H. Strauss. 2000. High Levels of Roundup® and Leaf-Beetle Resistance in Genetically Engineered Hybrid Cottonwoods. In *Hybrid Poplars in the Pacific Northwest:*

*Culture, Commerce and Capability*, edited by K.A. Blatner, J.D. Johnson, and D.M. Baumgartner. Pullman, WA: Washington State University Cooperative Extension Bulletin MISC0272, 29–38.

Meilan, R., A. Brunner, J. Skinner, and S. Strauss. 2001a. Modification of Flowering in Transgenic Trees. In *Molecular Breeding of Woody Plants, Progress in Biotechnology Series*, edited by A. Komamine, and N. Morohoshi. Amsterdam, Netherlands: Elsevier Science BV, 247–256.

Meilan, R., D.J. Auerbach, C. Ma, S.P. DiFazio, and S.H. Strauss. 2001b. Stability of Herbicide Resistance and GUS Expression in Transgenic Hybrid Poplars (*Populus* sp.) during Several Years of Field Trials and Vegetative Propagation. *HortScience* 37 (20): 1–4.

Meilan, R., K.-H. Han, C. Ma, S.P. DiFazio, J.A. Eaton, E. Hoien, B.J. Stanton, R.P. Crockett, M.L. Taylor, R.R. James, J.S. Skinner, L. Jouanin, G. Pilate, and S.H. Strauss. 2002. The *CP4* Transgene Provides High Levels of Tolerance to Roundup® Herbicide in Field-Grown Hybrid Poplars. *Canadian Journal of Forest Research* 32: 967–976.

Molkentin, J.D., B.L. Black, J.F. Martin, and E.N. Olson. 1996. Mutational Analysis of the DNA Binding, Dimerization, and Transcriptional Activation Domains of *MEF2C*. *Molecular and Cellular Biology* 16: 2627–2636.

Ostry, M., W. Hackett, C. Michler, R. Serres, and B. McCown. 1994. Influence of Regeneration Method and Tissue Source on the Frequency of Somaclonal Variation in *Populus* to Infection by *Septoria musiva*. *Plant Science* 97: 209–215.

Ostry, M.E., and K.T. Ward. 2003. Field Performance of *Populus* Expressing Somaclonal Variation in Resistance to *Septoria musiva*. *Plant Science* 164: 1–8.

Pilate, G., E. Guiney, M. Petit-Conil, C. Lapierre, J.-C. Leplé, B. Pollet, I. Mila, C. Halpin, L. Jouanin, W. Boerjan, W. Schuch, and D. Cornu. 2002. Field and Pulping Performance of Transgenic Trees with Altered Lignification. *Nature Biotechnology* 20: 607–612.

Rottmann, W.H., R. Meilan, L.A. Sheppard, A.M. Brunner, J.S. Skinner, C. Ma, S. Cheng, L. Jouanin, G. Pilate, and S.H. Strauss. 2000. Diverse Effects of Overexpression of *LEAFY* and *PTLF*, a Poplar (*Populus*) Homolog of *LEAFY/FLORICAULA*, in Transgenic Poplar and *Arabidopsis*. *Plant Journal* 22: 235–246.

Rugh, C.L., R.B. Senecoff, R.B. Meagher, and S.A. Merkle. 1998. Development of Transgenic Yellow Poplar for Mercury Phytoremediation. *Nature Biotechnology* 16: 925–928.

Serres, R., M. Ostry, B. McCown, and D. Skilling. 1991. Somaclonal Variation in Populus Hybrids Regenerated from Protoplast Culture. In: *Woody Plant Biotechnology*, edited by M.R. Ahuja. New York: Plenum, 59–62.

Shang, T.Q., S.L. Doty, A.M. Wilson, W.N. Howald, and M.P. Gordon. 2001. Trichloroethylene Oxidative Metabolism in Plants: The Trichloroethanol Pathway. *Phytochemistry* 58(7): 1055–1065.

Shani, Z., I. Levi, S. Mansfield, L. Roiz, M. Dekel, and O. Shoseyov. 2001. Modulation of Wood Fibers and Studies on the Mode of Action of Cellulose-Binding Domains (CBDs). Paper presented at the Tree Biotechnology in the New Millennium IUFRO/Molecular Biology of Trees Meeting, July 22–27, Stevenson, WA.

Sheppard, L.A., A.M. Brunner, K.V. Krutovskii, W.H. Rottmann, J.S. Skinner, S.S. Vollmer, and S.H. Strauss. 2000. A *DEFICIENS* Homolog from the Dioecious Tree

*Populus trichocarpa* Is Expressed in Both Female and Male Floral Meristems of Its Two-Whorled, Unisexual Flowers. *Plant Physiology* 124: 627–639.

Spiker, S., and W.F. Thompson. 1996. Nuclear Matrix Attachment Regions and Transgene Expression in Plants. *Plant Physiology* 110: 15–21.

Strohm, M., L. Jouanin, K.J. Kunert, C. Pruvost, A. Polle, C.H. Foyer, and H. Rennenberg. 1995. Regulation of Glutathione Synthesis in Leaves of Transgenic Poplar (*Populus tremula* x *P. alba*) Overexpressing Glutathione Synthetase. *Plant Journal* 7: 141–145.

Tappeiner, J.C. 1969. Effect of Cone Production on Branch, Needle and Xylem Ring Growth of Sierra Nevada Douglas-Fir. *Forest Science* 15: 171–174.

Tsai, C.-J., S.A. Harding, P. Pechter, R.J. Steinbrecher, D. Richter, and V. Chiang. 2001. Transgenic Aspen with Altered Lignin. Paper presented at the Tree Biotechnology in the New Millennium IUFRO/Molecular Biology of Trees Meeting, July 22–27, Stevenson, WA.

van Blokland, R., M. Ten Lohuis, and P. Meyer. 1997. Condensation of Chromatin in Transcriptional Regions of an Inactivated Plant Transgene: Evidence for an Active Role of Transcription in Gene Silencing. *Molecular and General Genetics* 257: 1–13.

Voinnet, O. 2001. RNA Silencing as a Plant Immune System against Viruses. *Trends in Genetics* 17(8): 449–459.

Weigel, D., and O. Nilsson. 1995. A Developmental Switch Sufficient for Flowering Initiation in Diverse Plants. *Nature* 12: 495–500.

# 5

# Exotic Pines and Eucalypts
## *Perspectives on Risks of Transgenic Plantations*

ROWLAND D. BURDON AND CHRISTIAN WALTER

Exotic pine and eucalypt species account for much of modern plantation forestry. Very large areas are involved, and cultivation and production are often very intensive. The use of genetic modification (GM), or genetic engineering, while still only in the development and testing phases for forest trees, is widely seen among those involved in tree improvement as largely an incremental development in these intensive cultivation systems. The GM of exotic pines and eucalypts raises many of the issues that are related to the GM of forest plantation crops in general (cf. Mathews and Campbell 2000). Some risks of GM are largely shared with conventional breeding, but some other, potentially important risks are specific to one or the other.

To avoid misunderstanding, we review concepts of risk in the context of the use of transgenic forest cultivars. We then identify key features that, with respect to risk factors, differentiate exotic pine plantations from traditional crop plants and other forest trees, and eucalypt plantations from exotic pine plantations. Features that are in varying degree specific to pines include crop lifespan (relating to both the potential magnitude of certain risks and the need for stable, long-term expression of transgenes); natural silvics (largely favoring species monocultures); soil and climatic tolerances (effectively restricting feasible control over growing environments); wind pollination (causing problems of containment); and the associated outbreeding system. We will also draw key comparisons between pines and eucalypts. For eucalypts, these include some calls for advances in transformation methods more pressing than those for pines; insect pollination compared with wind pollination in pines; and crop lifespans often markedly shorter than those of pines. Among the deployment-related risks, we emphasize the potential ecological risks and those relating to crop vulnerability, although we also consider human health issues and cultural issues. Among the ecological risks, possible horizontal gene transfer receives special attention, and we consider the application of the pre-cautionary principle largely in relation to this. We compare use of transgenic forest cultivars with evolution and conventional breeding, as a basis for identi-fying certain novel components of risk. In some respects GM mimics acceler-ated evolution, but it entails special risks because it skips the typical testing of low-frequency alleles by natural selection. We review strategies for risk man-agement in relation to likely probability profiles of different risk categories, emphasizing crop vulnerability. Some risks can be addressed by direct testing and other active countermeasures; however, there are some low-probability but potentially very serious events that call for risk spread. Finally, we address the point that all technologies, including GM, have their own risks, which must be weighed against their prospective benefits.

## Concepts of Risk in the Transgenic Context

We define the general concept of risk as a function (roughly, a product) of the probability of a negative effect occurring and its magnitude (or serious-ness). In turn, the magnitude can be viewed as roughly the product of the severity of the event and the scale on which it occurs (Burdon 1999); for instance, 100% mortality of a cultivar may be a severe event for that cultivar, but not a serious one if only a few trees have been planted. Where the proba-bility is unknown, the term "uncertainty" is widely used. Yet even if the probability is of a low but uncertain value, it cannot be readily discounted if the potential magnitude is extreme. In that situation, one must make judg-ments as to just how low the probability is, and there will be cases where there are good reasons for believing that it is essentially zero. The genetic engineering debate worldwide has become characterized by confusion over the term "risk." Although there are attempts to define and quantify risk, there is growing concern that the term is used misleadingly, with a readiness to view any nonzero probability as being effectively high and to ignore risks that arise without genetic engineering (Bazin and Lynch 1994). Moreover, the occurrence of an effect as such is often not clearly defined. The risk discus-sion is in many cases characterized by claims that a specific effect will occur, without reference to, and analysis of, the individual steps that must happen for it to occur.

A genetic transformation produces an organism that carries a new sequence of DNA, ideally at one, but often at more than one, location within the genome. If such a sequence already occurs in nature in one way or another, we have something akin to the possible outcome of natural gene transfer (also termed "horizontal gene transfer," or HGT). Although many opponents of genetic engineering evidently regard the possible HGT of artificially inserted DNA as a risk, and yet ignore the possible HGT of naturally occurring genes, no potential or perceived effect can actually be regarded as a risk without some analysis. A new sequence of DNA in an organism will often have an effect such as production of a new protein, if it codes for a structural gene, or, at the other extreme, simply incur the "wastage" of a small amount of resources to replicate the new DNA. The latter type of effect can be regarded as neutral with regard to the environment or human health. In some circumstances, the effect will benefit the organism, provided the new gene confers selective advantage. The effect may be negative if the new sequence of DNA reduces fitness. Also, an effect that benefits the organism might be a risk for the environment. The same, however, holds true for the outcome of a conventional breeding experiment, in which new gene combinations may be considered beneficial for some applications but a risk for the environment. The generation of an organism with unknown characteristics is a potential threat in relation to all human efforts to modify existing planting stock according to needs.

Any potential risk needs to be assessed in the context of the particular organism and in context of the environment the organism is living in. The engineered or bred product may have low-risk potential in a specific environment, but such may not be the case in other environments. Certain risks of crop failure can arise in plantations whether they result from conventional breeding or genetic engineering. Furthermore, a specific new genotype produced by either means must respond to the specific environment in which it is grown. The result of the effects of genotype and environment will finally decide the success or failure of the crop (genetically modified or not) and the potential risks.

## Distinguishing Features of Exotic Pine Crops

Crucial features that often distinguish pine plantations from other types of crop and other forest trees are crop lifespan (i.e., rotation age), natural silvics, soil tolerances, climatic tolerances, wind pollination, and the associated outbreeding system. Exotic status is a less consistent feature of pine plantations (given, for example, the huge areas of plantations of loblolly pine within its natural range), and its significance is more ambiguous in respect of biotic risks that may interact with the risks of genetic transformation (Burdon 1999). For example, an exotic crop may be growing in the absence of a disease, but if the disease arrives, and conventional breeding or transformation has accidentally (and perhaps without its being known) increased a susceptibility that is already high in the exotic environment, the outcome could be very serious. Invasiveness is often a natural feature of pines and other conifers (Rogers and Ledig 1996) that results from the dispersal potential of the winged seeds. It is not a result of artificial breeding or GM.

**Crop Lifespan.** Even though pines are often fast growing, they are seldom grown on very short rotations for reasons that include the following:

- They lack the extreme relative growth rates that allow mean annual increments to culminate at very early ages.
- Their wood is more valuable for solid-wood products, which often require relatively large logs, than for pulping or board products.
- Wood quality tends to be low, especially for higher-value markets, in very young trees.

The relatively long lifespan of pines can accentuate risks of any tree improvement technologies in several ways. For instance, a crop can be exposed to risk over a longer period. A delayed failure, which may still occur before profitable salvage is possible, will carry the compounded costs of crop establishment, tending, and protection. Furthermore, a delayed failure may involve plantings made over a number of years, which can be especially damaging in itself and preclude any rapid recovery of the forest-growing enterprise. On the other hand, if problems are observed during the initial years of establishment, further plantings can be stopped, whereas with annual crops vast plantings may be converted to transgenic types in a few years, as occurred in the United States with glyphosate-tolerant soybeans.

**Natural Silvics.** Pines are typically strongly light-demanding, like many species that are either pioneers in an ecological succession or are involved in fire-induced climaxes. That being the case, they are usually far more conveniently grown in pure, even-aged stands (Burdon 2002). Growing pines for convenience and high economic returns thus entails the risks that are inherent in growing pure, even-aged stands; however, the major part that pure, even-aged stands often play in the natural ecology is likely to reduce the inherent riskiness of the system.

**Soil and Climatic Tolerances.** Pines are generally adapted to relatively low soil fertility (Burdon 2002), with a number of important implications. Large areas of land are available, and that affects the potential scale of risk. The low fertility demands are likely to be related to the limitations in early growth potential, which affect crop lifespan. Moreover, the land available for pine plantations will often be sites where intensive intervention to counter some risk factors (e.g., certain diseases) is not economically feasible. The climatic tolerances of many pines also favor their use on sites imposing such constraints.

**Wind Pollination.** The wind pollination of pines is significant both because of the potential for long-distance spread of genetic material into exotic and native stands (e.g., Sedgley and Griffin 1989; DiFazio et al. 1999) and because of its likely impact on the level of diversion of resources into reproduction. Although most of the pollen settles within a short distance of its point of release, dense pollen clouds can occur at considerable distances from large areas of stand (Lindgren et al. 1995). If the species are exotics, ecological and political concerns over gene flow resulting from pollen contamination are less than they are with species that are native to the growing region, especially if none of the native species is interfertile with the plantation species. In fact, there are relatively few situations where such interfertility is likely to arise with pines.

The diversion of resources into reproduction can be major, and the pollen component can be very high in wind-pollinated species (e.g., Fielding 1960; Cremer 1992). This is all the more significant because the pollen component is "high-grade" biomass, in that it has a high content of nutrients.

Suppression of reproduction is an attractive goal, offering up to 100% reduction in pollen flow and a redirection of energy potentially resulting in increased timber production (Ledig and Linzer 1978), although costs of reproduction can be hard to establish experimentally (cf. Obeso 2002). Genetic engineering strategies exist that can largely achieve this goal, including the expression of cytotoxic genes in reproductive tissue, influencing reproductive pathways through homeotic genes, and the suppression of specific genes involved in reproductive development (Strauss et al. 1995). None of those, however, can guarantee zero reproduction when used alone, and a "pyramiding" strategy may be more successful in completely controlling reproduction in a given stand. This is of particular importance where interfertility with native species can be expected—if, indeed, the introgression of a genetically engineered DNA sequence is actually harmful.

Other, more subtle considerations can arise. For instance, elimination of pollen cone formation in pines may change crown configuration by eliminating zones of shoot without foliage. Thus the genotypes with the best crown configuration for crop productivity before suppression of pollen cone formation may not be so after suppression has been achieved. Hence there can be complex interactions between genetic engineering and classical breeding that could lead the unwary into unexpected losses. Complex interactions may arise among traits being addressed by conventional breeding, although some argue that they may be far less important because the changes that conventional breeding will bring about within a short time frame are less radical. That conclusion, however, still lacks supporting data, and more research is required to fully understand this issue.

Another environmental aspect related to sterility considerations is the importance of pollen-feeding native insects and birds (Mullin and Bertrand 1998). It may be appropriate to design strategies that allow production of nonviable pollen so that the normal food chain in a given ecosystem remains intact.

**Outbreeding Behavior.** Almost all pines are natural outbreeders (Richardson 1998; Burdon 2002), and that has important implications. Tree-to-tree genetic variation can be important for population resistance to diseases (Thielges 1982), which may not always fit conveniently with the clonal systems that may be needed for use of GM. The tree-to-tree genetic variation that is typically associated with the outbreeding behavior is likely to make genetic gains readily available from recurrent selection within existing populations. At the same time, such genetic improvement can be a crucial platform for capitalizing on the improvements achievable through GM because the merit of a transformant will inevitably be limited by the general merit of the recipient genotype.

**Exotic Status.** Where the pine species is grown as an exotic, genetic contamination of natural stands is not usually an issue, although hybridization with some other pine species might very occasionally be. If ex situ gene resources are to be maintained, as potential sources of new germplasm that is unrelated to existing material in breeding populations, any pollen contamination from commercial stands is generally very unwelcome, especially if such stands are intensively improved genetically and of narrow genetic base (Burdon and Kumar 2003). If it comes from transgenic material it is likely to be even more so, if only because it could raise a whole new area of public concerns. Suppression of pollen production would be sought in the interests of improving net production as well as allaying concerns over containment.

**Inherent Invasiveness.** Conferring herbicide resistance by means of GM is widely seen as creating a weed potential in an agricultural crop plant or exacerbating the weed potential of wild, interfertile relatives. In pines, herbicide resistance might, by creating a selective advantage when the herbicide is used, accentuate natural invasiveness both by seed dispersal from GM plantations and by their pollinating existing wildings, even though GM might compromise other components of fitness in the wild. Suppression of reproduction through GM, however, would remove even the natural invasiveness.

# Comparisons of Eucalypts with Pines

Eucalypts can be both highly vulnerable to weed competition and very subject to defoliation by insects (Cromer and Eldridge 2000). The traits of resistance to herbicides and resistance to insect attack are both being widely pursued through GM: these are two areas where GM will tend to be particularly attractive for eucalypts (e.g., Edwards et al. 1995; Harcourt et al. 1995). Both types of modification have recently attracted some attention in pines as well (Bishop-Hurley et al. 2001; C. Walter, unpublished data).

The insect pollination of eucalypts, as opposed to wind pollination of pines, will tend to reduce the risks of long-distance gene flow, meaning that containment is inherently easier, creating less call for suppression of reproduction. Eucalypt plantations have, until quite recently, been almost entirely exotic, unlike some huge areas of intensively cultivated pine plantations (e.g., *Pinus taeda* and *P. elliottii*) within their native ranges. Potential gene flow into natural populations has therefore been far less of an issue. However a recent upsurge of establishment of eucalypt plantations in Australia is changing the picture (Cromer and Eldridge 2000; Barbour et al. 2002). The very short rotations on which eucalypts are often grown, usually for pulpwood or fuelwood, should in some ways mitigate the impact of crop failure. Moreover, the easy terrain on which eucalypts are very often grown and the associated site cultivation should help facilitate protective intervention against, say, a disease outbreak, whether or not it arises specifically in transgenics. Where short rotations may lead to harvesting before production of pollen and/or seed, risks of unwanted gene flow would be greatly mitigated.

Interfertility among eucalypt species is common, which favors cross-pollination among species. It may be accentuated in exotics because flowering seasons can be greatly altered by their new environments, creating overlapping flowering times between species, which can break down natural reproductive isolation. However, any impact of GM through interspecific pollination is likely to be confined to ex situ genetic resources growing very close to transgenic crops. The short rotations, seemingly a mitigating factor, may not be entirely so; where eucalypt plantations feed highly capitalized pulp mills the costs of supply disruption through crop failure could be very high, unless alternative pulpwood supplies are readily available. This is, however, a management problem that is widely associated with the use of biological systems for production processes, and risks are associated with conventionally produced material and transgenic material alike. Moreover, in such cases the plantings made in just a single year, all of which might be exposed to some unsuspected risk factor, would represent a considerable fraction of a plantation estate.

## Risk Categories

The risks may be classified in various ways. The first breakdown that we adopt is the following: *development-related* risks, or those involved in pursuing and developing GM, and *deployment-related* risks, or those associated with the operational use of genetically modified crops.

A pervasive problem in evaluating both types is that many of the risks cited for GM have been greatly exaggerated. Some are largely unsubstantiated, and we believe that at least a large proportion of them will not be genuine risks at all. Although much has been written speculatively about potentially devastating effects from GM, some of the risks associated with the transgenic cultivars are largely shared with the products of conventional breeding and even of natural evolution.

## Development-Related Risks

As with any new or very immature technology, GM has risks relating to whether it will actually succeed, at least without prohibitive development costs (Burdon 1992). A spread of risks, in the genetic modifications that are pursued, may be indicated. That may entail much dispersal of effort; however, risk management is an expected component of applying any technology (*Australian/New Zealand Standard* 1999).

**Technological Risks.** Risk related to other forest plantation technologies should be reconsidered and analyzed in context and in comparison with new technologies. The fundamental science related to this area has been much neglected over the last decades, even though both clonal propagation and conventional breeding have generated examples of adverse effects (consider, for instance, somaclonal variation).

**Management Risks.** Those in the field of tree breeding tend to be very concerned that pursuit of GM will be at the expense of appropriate effort on other, classical technologies (although substitution of inputs may be reduced by the amount of resources that would be forthcoming for GM but not the other technologies). The appropriate allocation of resources to the different technologies is a key management issue (Burdon 1992; 1994). Competition for research funding, combined with stagnant funding levels, often means that only lip service is paid to the complementarities between new biotechnology, of which GM is a part, and classical field-based breeding (Burdon 1992). Yet GM represents an intensification of the domestication process, which is fundamentally about making increased inputs into growing crops to improve returns (cf. Burdon 1994). There are strong perceptions among tree breeders that there is a major substitution of effort rather than a boosting of total effort on tree improvement. Many breeders believe that if GM is pursued significantly at the expense of classical breeding, with its infrastructure of progeny plantings and gene resources, much damage could be done. Transformation scientists, however, may not share that concern.

An associated problem is staffing and allocation of personnel to decision-making roles. The challenge is to combine leadership in the new technology with a strategic vision, so as to integrate biotechnology with the field-based components of genetic management. For both the breeding and the GM camps this challenge is important.

## Deployment-Related Risks

Briefly, deployment-related risks may be grouped into the following, partly overlapping categories:

- ecological risks;
- human health risks;
- cultural and religious issues; and
- crop vulnerability.

**Ecological Risks.** Ecological risks involve, in principle, two avenues of gene flow:

- pollen flow into natural populations, invasion of natural ecosystems through seed produced by transgenic crops, and transgenic cultivars' developing a weed potential in their own right (FAO 2002); and
- spontaneous horizontal gene transfer (HGT) in the field (Syvanen 1994; Dale 1999; Mullin and Bertrand 1998).

All such risks depend heavily on the introduced gene(s) conferring a material fitness advantage in the wild. Often there is no reason to expect any such advantage, and any effect must always be evaluated in context with the particular environment the organism is placed in. Any effect and its magnitude must also be compared with those arising from the use of already accepted and practiced tree improvement technologies. Furthermore, if HGT occurs at reasonable frequency, its impact may not be any greater than that of the simultaneously occurring HGT of identical genes in a natural environment (Jain et al. 1999).

Among the perceived risks for transgenic forest plantations, HGT is perhaps the most difficult to discuss on a rational, properly informed basis. HGT also poses the problems of not being amenable to risk spread and of potentially being irreversible if it occurs. It is a well-known phenomenon in nature, particularly between bacterial species (Lorenz and Wackernagel 1994; Eisen 2000), and HGT into and between higher organisms has also been postulated and occasionally demonstrated (Nielsen et al. 1998; Kado 1998). The transfer of specific DNA sequences from *Agrobacterium* sp. to cells of plant species during an infection process can also be regarded as HGT. This is currently the only known example of a type of HGT that not only occurs frequently in nature but also involves the transfer of DNA from a microorganism to a higher organism. Viruses may also play a role in gene transfer between higher species, but this issue is much less researched. The transfer of DNA from gut bacteria to cells lining the mammalian gut has occasionally been postulated, but solid scientific data to support this hypothesis have not been produced. The study of HGT so far has led to the notions that the total genome, including all organisms, may in fact be highly flexible (Jain et al.

1999; Lawrence 1999) and that HGT is a common tool of evolution (de la Cruz and Davies 2000; Woese 2000). It is therefore highly questionable whether GM would add materially to the effects of natural mutation combined with any natural capacity for HGT, given that mutation and recombination surely occur frequently among the vast numbers of individual microorganisms.

For HGT to cause harm, all of four conditions must be met:

1. HGT first must occur.
2. The new sequence must be integrated into the host genome and successfully expressed.
3. The resulting gene product must then confer a selective advantage on the recipient organism.
4. That selective advantage must then result in ecological harm.

Looking at the situation slightly differently, the probability of harm arising from HGT represents the probability that HGT will occur and the new gene establish itself (Steps 1 and 2), multiplied by the probability that harm will then result (Steps 3 and 4). If the probability of occurrence is high, that will almost certainly indicate that HGT occurs regularly but, even in conjunction with ubiquitous mutation in microorganisms, seldom if ever does any ecological harm. Thus if there is a high probability of HGT occurring, it seems very unlikely to be ecologically harmful if it does occur.

**Human Health Risks.** This category is technically problematic, generally representing tenuous possibilities, even with most food crops, although vehement expressions of concern are now often heard (Ho 1998; Antoniou 1996). The only foodstuffs that pines provide directly are seeds, and the species concerned are not major plantation crops. Indirect food production occurs through the collection of fruiting bodies of edible symbiont fungi, which again raises the issue of HGT (Droege et al. 1998). Other, very hypothetical possibilities arise through the role that pine material might play in natural food chains or in incidental contamination, for example, through deposition of pollen on food crops. However, if a gene transformed into pine could contaminate food sources, it is likely that it could also do so in flowing from its original, natural host to potential foods. The difference between the two sources appears significant to opponents of genetic engineering, but we are not aware of any data to substantiate their claims.

With issues of human health and ecological side effects, many parts of society call for applying the precautionary principle, which places the onus of proof on the safety of a new technology. On the other hand, many of the undesirable possibilities invoked seem remote indeed and not substantiated by scientific data. Concerning foods and food chains, even secondary products of the very small number of new genes that would be used in GM would arise in the context of many thousands of natural products that are routinely detoxified in the small amounts that occur (unless food spoilage is involved, such as the production of large amounts of aflatoxin). Furthermore, the application of the precautionary principle in the way that some opponents of modern biotechnologies advocate could potentially prohibit any new human activity that is based on a new technology, thereby stifling almost all progress.

**Cultural and Religious Issues.** It has become evident that significant parts of society regard the transfer of genes from one organism to another as contrary to their religious or cultural beliefs and therefore unacceptable. Such beliefs, whatever their origins, are often extremely difficult to address and overcome, because they can seal people off from the processes of education and open-minded discussion of the facts about, and foreseeable consequences of, using transgenics. We dispute the merit of those objections. Also, we contend that genetic engineering needs to be reviewed critically in comparison with older technologies, such as breeding and the mass clonal propagation that have already been widely used to deliver genetic gains.

**Crop Vulnerability.** We consider the little-publicized issue of crop vulnerability an important one, and we will give it special attention. Its economic significance and hence its social significance are potentially enormous.

In Table 5-1 we summarize the categories of technical risk, which are primarily associated with deployment of transgenics, along with risk properties, predisposing risk factors, and appropriate types of countermeasures. Putative risk factors relating to possible cultivar decline associated with genetic transformation are summarized in Table 5-2. The two tables provide a backdrop for much of the following discussion.

Crop vulnerability may arise from GM in various ways (Table 5-2), some direct and some indirect. The more direct ways involve side effects of the action of new structural genes or of modification (either overexpression or down-regulation) of the action of existing genes; side effects of the process of inserting the new DNA sequences; or silencing of newly introduced genes (Finnegan and McElroy 1994; Matzke and Matzke 1995). Similar effects may arise from sets of new genes that, in combination, have the potential to interact with each other (Burdon 1999) to create undesirable effects. Indirect effects can arise, for instance, from the impact on deployment practices that would follow from the use of transgenic material, for example, a shift toward use of clonal material, or if the environmental range of a species is extended after a disease resistance gene has been inserted and the resistance later breaks down through pathogen mutation (Burdon 1999). However, strategies to delay the evolution of pest resistance have recently been discussed (Cohen and Gould 2000).

**TABLE 5-1.** Categories of Technical Risk, Putative Risk Properties, and Risk Factors, with Potential Approaches to Countering Them

| Risk category | | Likelihood | Potential severity | Predisposing factors | Prime countermeasures |
|---|---|---|---|---|---|
| General | Specific | | | | |
| Related to technology development | Transient gene expression | High | Troublesome, but in some situations could even be advantageous | Transformation technique; possibly gene(s) concerned | Careful testing |
| Ecological | Direct contamination of ecosystems | Largely precluded by exotic status of genera in question | Problem in managing ex situ gene resources | Wind pollination (species native to area, presence of interfertile relatives), efficient seed dispersal | Conferring sterility |
| | Horizontal transfer | Frequent among micro-organisms; unlikely for higher organisms | Most unlikely to be significant | Highly dependent on selective advantage of transgene(s) in field | HGT mechanism needs to be better understood; more research needed |
| Human health | Allergenicity | Extremely low for transgenes | Most unlikely to be serious | Wind pollination; quest for durable heartwood | No specific measures |
| | Food contamination | Extremely low | Most unlikely to be serious | Conceivably applicable to nectar (eucalypts) or edible fungi (pines) | No specific measures envisaged |
| Crop vulnerability | Cultivar decline/failure | General hazard; widely variable | Very high; at top of range (see Table 5-2) | Various (see Table 5-2) | Various (see Table 5-2) |
| | Nondurability of resistance | Potentially significant | Potentially troublesome | Reliance on single genes of large effect for resistance | Using multiple resistance factors (with its own risks) |

*Note:* Risk categories need not be mutually exclusive. For example, horizontal transfer could theoretically lead to food contamination, whereas nondurability of resistance could cause cultivar decline/failure.

**TABLE 5-2.** Factors Believed to Generate Risks of Cultivar Decline or Failure

| Factor | *Ranking of risk potential for categories within the factor (descending order)* | *Preferred countermeasures and/or remarks* |
|---|---|---|
| Transgene source | Synthetic genes > genes from distant taxa > genes from close relatives > genes from within species[a] | Risk spread a potential defense apart from choice of source. |
| Transformation method | Biolistics > *Agrobacterium* | Risk spread potentially very effective defense; *Agrobacterium* poses greater technical difficulty with taxa concerned. |
| Role of transgenes | Structural genes > regulators or anti-sense sequences | Lower rankings unclear. |
| Type of genes | Homeotic genes (e.g., flowering) > other genes | May constitute a key hurdle in suppressing reproduction. |
| Magnitude of gene effect | Large > small | Major genes preferred nonetheless, for various reasons. |
| Number of gene insertions | Multiple > few > single | Multiple insertions may be needed on regulatory grounds or to confer durability of resistance. |
| In vitro culture technique | Single-cell lines > organogenetic cultures Adventitious shoots > axillary shoots | |

*Note:* Many of the conclusions here must be based on opinion, for lack of solid data.
[a]Condition favoring use of traditional breeding.

   The corn blight epidemic in the United States in 1970 (Levings 1989) may be the classic object lesson. It resulted from massive reliance on the Texas cytoplasmic male-sterility factor to produce the hybrid maize, an unsuspected side effect of which turned out to be extreme susceptibility to a new strain of the pathogen, and the eventual appearance of that pathogen strain. Admittedly, that case involved a mutant gene in an organelle genome, and it did not involve GM, so it is not strictly parallel to what would be achieved by genetic transformation involving the nuclear genome. How relevant the case is to genetic transformation in general is a matter of opinion. Also in the realm of opinion is whether, if it is relevant, it is so to a broad spectrum of genetic transformations or specifically to ones involving flowering. One breeding scientist's view is that it is prudent to assume the former, even though the homeotic nature of various flowering genes gives reason to believe that risks might be greater when suppression of flowering is involved. This case can also be seen as an illustration of the fact that unwanted, unexpected, and sometimes hazardous effects do not occur only from the use of material produced by GM. Classical breeding efforts can produce them as well, although the case in point, while arising in the era of classical breeding, was achieving something that is now being pursued by GM. Any transgenes that are used, and their "downstream" products, will admittedly be much better characterized than this male-sterility factor, but that will not eliminate all possibility of nasty surprises. In conclusion, a case-by-case analysis of any modification, whether developed by genetic engineering or some other technology, appears mandatory to mitigate risk. Since zero risk is essentially unattainable with any human activity, risk spread must almost always be considered, and it is a topic to which we will return.

**The Precautionary Principle.** This is an area of major uncertainty. What is an appropriate level of the proof of safety (Van den Belt 2003)? There is also scope for enormous variation in subjective assessments of the hazards, particularly hazards relating to ecological side effects and human health. At one extreme, some will see virtually limitless need to apply the precautionary principle. They will tend to argue that, although individual hazards are slight, they are not so numerous as to generate a significant total hazard. Those who take this position tend to be exasperated by what they see as the uncritical enthusiasm of the optimists. At the other extreme, some people will see all the hazards as being very minor. This group may include devotees of James Lovelock's Gaia principle, whereby a biota and its interdependent environment have an enormous inherent resilience. To many of the optimists, an extreme adoption of the precautionary principle is something like allegations of child abuse in an acrimonious custody battle, which are easily made but inherently very hard to disprove. Interestingly, there is no commonly agreed "precautionary principle," as many different variations exist (Smith 2002). Indeed, in connection with the recent Royal Commission on Genetic Engineering in New Zealand, no fewer than 35 different versions of the principle were found among legal texts (Christensen 2000).

A major societal issue related to genetic engineering of all crops is the lengths to which some partisans apparently will go to in trying to impose viewpoints that, while they may be dearly held, seem most improbably alarmist in the context of all available evidence. Many claims are made based on lack of scientific understanding or even refusal to accept a rigorous scientific approach to the issue. Very often doomsday scenarios are presented, with little or no scientific evidence to support them. This was particularly apparent in submissions to the aforementioned Royal Commission on

Genetic Engineering in New Zealand, where proponents and opponents of genetic engineering could present their cases (www.gmcommission.govt.nz). For example, a witness brief by Dr. Elaine Ingham (then of Oregon State University affiliations), who testified for the Green Party of New Zealand, cited to the commission a nonexistent paper to support a claim that genetically modified *Klebsiella planticola* bacteria had, if released, the potential to devastate plant life on the planet. When the nonexistence of the paper was exposed (Walter et al. 2001), Dr. Ingham and the Green Party of New Zealand had to apologize to the commission members for misleading them (Fletcher 2001). Furthermore, the use of data by Dr. Ingham to support her case before the commission has proved to be scientifically unsustainable (Walter et al. 2001). Ingham's conclusions about the potentially harmful nature of the bacterium were based on a single experiment in the laboratory, where controls were not properly applied. Also very important is that a natural variant of *Klebsiella planticola* shares the key, alcohol-producing characteristic of the genetically engineered bacterium (Jarvis et al. 1997) without showing any harmful effect on plant life in its natural environment.

### Low-Risk Applications

Uses of genetic transformation will not be confined to operational ones in commercial crops, but show great promise as a research tool. Transgenes can be used to study pathways of gene action and their effects on phenotype, as an aid to identifying appropriate breeding goals, and to help identify desirable genes that might be exploited by enhancements of classical breeding. Even with purely research-oriented applications, some may still see risks, notably in the area of containment, but the magnitude of such risks should be far less than for those associated with operational use and deployment.

## Genetic Modification Compared with Evolution and Classical Breeding

New genes have appeared and spread through populations throughout evolutionary history, even though certain genes that perform basic functions may be highly conserved. It is evident that various processes that allow the transfer of DNA sequences from one organism to another, such as HGT between bacteria, *Agrobacterium* gene transfer to plant cells, and virus infections, have all contributed to variation and long-term evolution (see above). It is instructive, though, to compare the process of introducing a transgene with these natural processes whereby a gene can spread through a population.

In nature, a new allele will typically appear as a rare variant. It can have various fates, ranging from rapid, often random extinction, through persistence at low if variable frequencies, to spreading through the population and ultimately becoming fixed. The exact fate will depend whether it is selectively neutral, or nearly so, on the one hand, or significantly advantageous, on the other, plus large elements of chance. Alternatively, new alleles can enter the population by migration (usually pollen and/or seed dispersal). Either way, a new allele tends to enter the population on a "toe-in-the-water" basis. If such a gene proves deleterious, even with some time lag, it will be eliminated or remain at a low frequency with minimal cost in population fitness. Population bottlenecks of one sort or another can accelerate the process, and even set evolution on new courses, but will be accompanied by risks of extinction. Yet even a population bottleneck would seldom, by itself, bring a gene to a frequency that would immediately allow it to pervade the population. Viruses have the potential to spread genes rapidly through populations of many organisms and even across species barriers. However, viruses for pines or eucalypts have not yet been described.

Use of a transgene, by contrast, can immediately raise the effective frequency of a new gene to 100% in a portion of the crop's range, if its action is fully dominant, which would create 100% exposure to any adverse side effects. However, this holds only for the crop in question. A transfer of the gene to another organism by HGT would only occur at very low frequency, and therefore the frequency of this gene in the recipient organism would also be very low, essentially as would be the case with a fresh mutation. Clearly, though, there are various categories of transgenes, ranging from constructs that modify the expression of existing structural genes, which are likely to present relatively low risks, to structural genes of totally novel function within the species, for which the attendant risks may be much higher (Table 5-2). Although HGT may lead to rapid spread of genes through populations of some organisms, it appears most unlikely to have any serious impact on a commercial plantation or its associated biota.

Artificial selection is in several respects intermediate between GM and natural evolution. In that it depends strictly on genes that have occurred naturally within living populations of a given organism it stands close to evolution. However, the intensive and highly directional selection that it can entail can generate closer parallels with use of transgenics, especially if technology can be used to select for specific genes that have major phenotypic effects.

Operational use of transgenics can thus lead to more rapid and complete exposure to risks of adverse side effects of specific genes than can be expected in natural evolution or even classical artificial breeding, in which massive substitutions of particular alleles will tend to be far slower (except for high-risk cases of some complete clonal monocultures). This does not mean that undue risk exposure cannot be incurred with conventional breeding, particularly when combined with mass vegetative propagation for clonal forestry (e.g., Libby 1982; Burdon 2001), but it does mean an essentially new area of risk that poses its own management challenges.

Classical breeding has important risks of its own (Burdon 1999). It may be embarked upon in a wrong species or a wrong population, or it may address a wrong breeding goal, any of which can totally negate genetic gains. Such mistakes can be exacerbated by the long generation intervals for breeding and by difficulties and delays in developing the necessary mass propagation technology. Adverse genetic correlations, if not identified, can, along with market uncertainties, lead to misdefinition of breeding goals. Careless narrowing of the genetic base, which is a temptation when rapid genetic advances are being made and mass multiplication technology is available, can easily lead to crop vulnerability.

## Risk Profiles and Risk Management

Our focus is on deployment-related risks (Table 5-1). Any risk will have a profile of probability in relation to severity, as outlined at the beginning of this chapter. In principle, this will conform to a mathematical function but one that is seldom closely defined; typically there will be a high probability of minor losses, trailing away to much lower probabilities of much higher percentage losses. In fact, we are typically looking at low probabilities of disastrous outcomes, albeit with almost no idea of exactly how low those probabilities are, although some transformations appear less risky than others (Table 5-2). However, the length of rotation can contribute to the severity of such disasters, depending on the exposure to specific risks. The potential seriousness, even if the probability is low, demands some form of risk management (Burdon 1999; 2001). If active countermeasures against risks cannot be well targeted, but risk spread is feasible, then risk spread is strongly indicated.

In practice, we will be working with very little quantitative information—or, in the jargon, with great uncertainty. True, there are many cases where transformation has been associated, directly or indirectly, with disastrous effects on fitness, but they will often be manifested in cell lines that are obviously not worth committing to field trials, although the exact reasons for their state are seldom identified. For the material with enough promise to be tested in the field, the risks, in terms of the probability function in relation to severity, remain very uncertain. However, the potential seriousness, which may be extreme with delayed manifestation of induced disease susceptibility, may dominate the appropriate risk management. A significant period of field testing (which can be preceded by laboratory testing and physiological studies) seems essential in some, but not all, cases to eliminate effects such as greatly enhanced susceptibility to climatic damage (for example, in a transformant for wood chemistry that might adversely affect the mechanical stability of the tree). However, such testing will only reduce the risks, not totally eliminate them, and in the short term is liable to greatly erode the potential time savings in using GM compared with conventional breeding. The transformation scientist, however, looks to the future when, with the risks of transgenes being much better known, the period of field testing can be shortened to allow major time savings from using GM compared with conventional breeding.

In general, society must make decisions on using this new technology, and that will involve weighing the prospective benefits against the risks. There will be considerations of how the risks can be addressed, whether by risk spread or active countermeasures, or both (Table 5-1; see also Burdon 2001), and what remaining level of risk may have to be accepted if the technology is to be used. This is true, however, for any technology in use, and people will always accept some risk, as long as sufficient benefits are realized or anticipated. Determination of the amount and level of field testing of transgenic material needed before commercial release has to be guided by scientific argument but also by the level of residual risk a society is prepared to take.

With respect to the use of transgenic plantations, we consider risk spread a key component of risk management. It appears especially appropriate for addressing potentially very serious events of low but very uncertain probabilities. An ideal risk-spread strategy for GM or conventional products should be based on the following planks (cf. Burdon 1999):

1. As a starting point, a diversity of recipient genotypes should be used.
2. Creation of transgenic plantations should involve different gene-insertion events, and the insertion should be as precise as possible.
3. A number of different transgenes or regulator constructs should be used in parallel, to achieve a particular objective,
4. The process should include measures to avoid, or at least minimize, the cotransfer of unwanted DNA sequences.

Requirement 3 may be difficult to meet with GM designed to eliminate all reproductive activity, which would be desired for reliable genetic containment as well as for optimizing resource allocation for crop production. However, opinions may vary on the need for complete suppression and, if suppression requirements are less stringent, how to meet them in cases where a number of independent suppressor genes could be available in various combinations.

Meeting Requirement 3 may, ironically, be impeded by regulatory mechanisms that require processing of all elements of risk spread separately, which increases the costs of compliance, rather than addressing an appropriate spread of risks as a single package.

For very-low-probability events, such risk spread, where it is applicable, can reduce the probability of catastrophic crop failure by orders of magnitude. In other words, the probability of simultaneous disasters, each of low probability, involving an unacceptable proportion of a set of independent transformations should become almost vanishingly low. On the other hand, such use of multiple transgenes will increase the theoretical risks of HGT, but only by a factor of the number of risk-spread elements, assuming that no such element can be identified as incurring elevated HGT-related risks. Thus the relative reduction in risks of GM-related crop failure should far exceed the relative increase in HGT-related risks.

For the longer term, leading into active countermeasures against the risks, more research into gene integration and expression characteristics, as well as on the influence of transgenes on natural ecosystems, should help to define constructs that carry a significantly lower risk potential. Examples are new strategies for the selection of transgenes avoiding antibiotic selection or the complete elimination of selective markers (Joersbo and Okkels 1996; Haldrup et al. 1998; Sugita et al. 1999). Although a perceived negative effect of antibiotic-resistance markers is not substantiated by scientific data, and one can involve antibiotics that are of no further clinical interest, the avoidance of such markers may increase public acceptance of genetic engineering.

There may be a call to use multiple transgenes simultaneously in cultivar genotypes, based on technical needs and/or regulatory requirements for genetic containment in field deployment. This has a potential, which is admittedly very speculative, to increase greatly the inherent risks associated with deployment of transgenic material (Burdon 1999). Not only will there be a simple accumulation of the risks associated with each individual transgene, but increasing numbers of transgenes will entail greatly increasing numbers of combinations that might generate adverse interactions between different transgenes if the transgenes involve significant risk factors (Table 5-2). Yet the use of multiple transgenes should actually reduce certain deployment-related risks by providing safeguards against the operational failure of one of the transgenes (Cohen and Gould 2000).

## Conclusion

In this chapter we have focused largely on the risks associated with genetic transformation. The risks, however, need always to be balanced against the prospective benefits, which will require critical examination with respect to both their validity and their dependence on the use of genetic transformation. The risks must also be placed in perspective with those of alternative methods of achieving the same goals; for example, conventional breeding, which will have its own risks, although these are currently much more readily accepted by society. Moreover, GM can be used as a research tool largely without incurring the risks attendant upon operational use in commercial crops.

A breeding scientist may argue that the risks associated with conventional breeding can be more easily managed than those of transgenics, which typically involve new genes of large effect. However, a transformation scientist may consider the less-predictable risks of conventional breeding, which are associated with many unidentified genes, to be greater than those of GM involving one characterized gene. The comparative risk potentials of the alternative technologies remain arguable. Possibly much more important is that GM can in some cases lead to more rapid and complete exposure to certain risks. With some products of conventional breeding, however, this problem can also arise.

Experience with other technologies, notably biotechnology, is that nature often springs nasty surprises, just when problems seem to be solved. Resistance to antibiotics, the dangers associated with blood transfusions, and the corn blight epidemic of 1970 are just three obvious examples. They illustrate, however, the need for a comprehensive risk/benefit analysis and an informed decision on what level of risk we can accept in return for obtaining a certain level of benefit, in a context of an unavoidable element of uncertainty. Notwithstanding the problems of antibiotic resistance, no one would seriously demand that we stop using a key weapon for combating many of the worst human diseases. The benefits of genetic engineering are becoming obvious in many areas. In medicine they are huge and are often very widely accepted. In agriculture they are also great but are now beset by problems of public acceptance. The example of Golden Rice, a genetically modified and vitamin A-enhanced rice that has the potential to save hundreds of thousands from blindness, comes to mind. In forestry the prospective benefits are also great (Strauss et al. 1999; Fenning and Gershenzon 2002), but the ecological ramifications pose problems of acceptance, and the time frames for plantation crops both create attractions for use of GM and accentuate certain risks. The lesson is surely that those who manage the applications of technologies like GM need good advance preparations for surprises, and they need management tools to minimize any impact on the environment, human health, or economic prosperity.

We can make our own technical judgments of risks, enhance those judgments with quantitative risk analysis, and devise risk-management strategies. Yet we cannot ignore public perceptions, even if we believe those perceptions are at odds with the facts. Moreover, scientific fiascos, no matter how rare, can have strong and enduring effects on public confidence. Subjective judgments on the part of the public may be perceived vividly by scientists, but scientists can never avoid areas where they must be guided in some degree by their own subjective judgments.

# References

Antoniou, M. 1996. Genetic Pollution. *Nutritional Therapy Today* 6(4): 8–11.

*Australian/New Zealand Standard: Risk Management.* AS/NZS 4360:1999. Homebush, Australia: Standards Australia, and Wellington, New Zealand: Standards New Zealand.

Barbour, R.C., B.M. Potts, R.E. Vaillancourt, W.N. Tibbits, and R.J.E. Wiltshire. 2002. Gene Flow between Introduced and Native Species. *New Forests* 23(3): 177–191.

Bazin, Michael J., and James Michael Lynch, eds. 1994. *Environmental Gene Release. Models, Experiments and Risk Assessment.* Suffolk, UK: Chapman & Hall.

Bishop-Hurley, S.L.R., J. Zabkiewicz, L.J. Grace, R.C. Gardner, and C. Walter. 2001. Conifer Genetic Engineering: Transgenic *Pinus radiata* D. Don and *Picea abies* Karst Plants Are Resistant to the Herbicide Buster. *Plant Cell Reports* 20: 235–243.

Burdon, R.D. 1992. Tree Breeding and the New Biotechnology—In Damaging Conflict or Constructive Synergism? In *Proceedings, IUFRO Meeting on Breeding Tropical Trees*, Cartagena and Cali, Colombia, October 1992. CAMCORE, North Carolina State University, Raleigh NC, 1–7.

———. 1994. The Place of Biotechnology in Forest Tree Improvement. *Forest Genetic Resources* 22: 2–5 (Rome, Italy: FAO).

———. 1999. Risk-Management Issues for Genetically Engineered Forest Trees. *New Zealand Journal of Forestry Science* 29: 375–390.

———. 2001. Genetic Aspects of Risk—Species Diversification, Genetic Management and Genetic Engineering. *New Zealand Journal of Forestry* 45(4): 20–25.

———. 2002. An Introduction to Pines. In *Pines of Silvicultural Importance*, compiled by CAB International. Wallingford, UK: CAB International, x–xxi.

Burdon, R.D., and S. Kumar. 2003. Stochastic Modelling of the Impacts of Four Generations of Pollen Contamination in Unpedigreed Gene Resources. *Silvae Genetica* 52: 1–7.

Christensen, S. 2000. Personal communication with the authors.

Cohen, M.B., and F. Gould. 2000. *Bt* Rice: Practical Steps to Sustainable Use. *International Rice Research Notes* 25(2): 4–10.

Cremer, K.W. 1992. Relations between Reproductive Growth and Vegetative Growth in *Pinus radiata. Forest Ecology and Management* 32: 179–199.

Cromer, R.N., and K.G. Eldridge. 2000. The Eucalypts as Tree Crops. In *Ecosystems of the World, vol. 19: Tree Crop Ecosystems*, edited by F.T. Last. Amsterdam, Netherlands: Elsevier, 226–269.

Dale, P.J. 1999. Public Concerns over Transgenic Crops. *Genome Research* 9: 1159–1162.

De la Cruz, I., and I. Davies. 2000. Horizontal Gene Transfer and the Origin of Species: Lessons from Bacteria. *Trends in Microbiology* 8(3): 128–133.

DiFazio, S., F. Leonardi, W.T. Adams, and S.H. Strauss. 1999. Potential Impacts of Hybrid Poplar Plantations on Black Cottonwood Plantations (Abstr.). In *Proceedings of the Western Forest Genetics Association Meeting*, Flagstaff, AZ, July 1999.

Droege, W., A. Puehler, and W. Selbitschka. 1998. Horizontal Gene Transfer as a Biosafety Issue: A Natural Phenomenon of Public Concern. *Journal of Biotechnology* 64(1): 75–90.

Edwards, G.A., N.W. Fish, K.J. Fuell, M. Keil, J.G. Purse, and T.A. Wignall. 1995. Genetic Modification of Eucalypts—Objectives, Strategies and Progress. In *Eucalypt Plantations: Improving Fibre Yield and Quality.* Proceedings of the CRCTHF-IUFRO Conference, Hobart, Australia, February 19–24, 1995. Hobart: CRC for Temperate Hardwood Forestry, 389–391.

Eisen, J.A. 2000. Horizontal Gene Transfer among Microbial Genomes: New Insights from Complete Genome Analysis. *Current Opinion in Genetics and Development* 10(6): 606–611.

FAO (UN Food and Agriculture Organization). 2002. *Summary Document (Long Version) of FAO Electronic Conference "Gene Flow from GM to Non-GM Populations in the Crop, Forestry, Animal and Fishery Sectors."* Rome, Italy: FAO. http://www.fao.org/biotech/logs/C7/summary.htm (accessed November 12, 2002).

Fenning, T.M., and J. Gershenzon. 2002. Where Will the Wood Come From? Plantation Forests and Biotechnology. *Trends in Biotechnology* 20(7): 291–296.

Fielding, J.M. 1960. Branching and Flowering Characteristics of Monterey Pine. Canberra, Australia: Forestry and Timber Bureau, Bulletin No. 37.

Finnegan, J., and D. McElroy. 1994. Transgene Inactivation: Plants Fight Back! *Bio/Technology* 12: 883–888.

Fletcher, L. 2001. New Zealand GMO Debacle Undermines Green Lobby. *Nature Biotechnology* 19: 292.

Haldrup, A., S.G. Petersen, and F.T. Okkels. 1998. Positive Selection: A Plant Selection Principle Based on Xylose Isomerase, an Enzyme Used in the Food Industry. *Plant Cell Reports* 18: 76–81.

Harcourt, R., X. Zhu, D. Llewellyn, E. Dennis, and J. Peacock. 1995. Genetic Engineering for Insect Resistance in Temperate Plantation Eucalypts. In *Eucalypt Plantations: Improving Fibre Yield and Quality.* Proceedings of the CRCTHF-IUFRO Conference, Hobart, Australia, February 19–24, 1995. Hobart: CRC for Temperate Hardwood Forestry, 406–408.

Ho, M.W. 1998. *Genetic Engineering: Dream or Nightmare?* Bath, UK: Gateway.

Jain, R., M.C. Rivera, and J.A. Lake. 1999. Horizontal Gene Transfer among Genomes: The Complexity Hypothesis. *Proceedings of the National Academy of Sciences USA* 96(7): 3801–3806.

Jarvis, G.N., E.R.B. Moore, and J.H. Thiele. 1997. Formate and Ethanol are the Major Products of Glycerol Fermentation Produced by *Klebsiella planticola* Strain Isolated from Red Deer. *Journal of Applied Microbiology* 83: 166–174.

Joersbo, M., and F.T. Okkels. 1996. A Novel Principle for Selection of Transgenic Plant Cells: Positive Selection. *Plant Cell Reports* 16: 219–221.

Kado, C.I. 1998. Agrobacterium-Mediated Horizontal Gene Transfer. *Genetic Engineering* 20: 1–24.

Lawrence, J.G. 1999. Gene Transfer, Speciation, and the Evolution of Bacterial Genomes. *Current Opinion in Microbiology* 2(5): 519–523

Ledig, F.T., and D.I.H. Linzer. 1978. Fuel Crop Breeding. *Chemtech* 8: 18–27.

Levings, C.S. 1989. The Texas Cytoplasm of Maize: Cytoplasmic Male Sterility and Disease Susceptibility. *Science* 250: 940–947.

Libby, W.J. 1982. What Is a Safe Number of Clones per Plantation? In *Resistance to Diseases and Pests in Forest Trees*, edited by H.M. Heybroek, B.R. Stephan, and K.

von Weissenberg. Proceedings of the 3rd International Workshop on Genetics of Host-Pathogen Interactions in Forestry, Sept. 1980, Wageningen, Netherlands, 342–360.

Lindgren, D., L. Paule, X.H. Shen, R. Yazdani, U. Segerstron, J. Wallin, and M.L. Lejdebro. 1995. Can Viable Pollen Carry Scots Pine Genes over Long Distances? *Grana* 34: 64–69.

Lorenz, M.G., and W. Wackernagel. 1994. Bacterial Gene Transfer by Natural Genetic Transformation in the Environment. *Microbiological Reviews* 58(3): 563–602.

Mathews, J.H., and M.M. Campbell. 2000. The Advantages and Disadvantages of the Application of Genetic Engineering to Forest Trees: A Discussion. *Forestry* 73(4): 371–380.

Matzke, M.A., and A.J.M. Matzke. 1995. How and Why Do Plants Inactivate Transgenes? *Plant Physiology* 107: 679–685.

Mullin, T.J., and S. Bertrand. 1998. Environmental Release of Transgenic Trees in Canada—Potential Benefits and Assessment of Biosafety. *The Forestry Chronicle* 74(2): 203–219.

Nielsen, K.M., A.M. Bones, K. Smalla, and J.D. van Elsas. 1998. Horizontal Gene Transfer from Transgenic Plants to Terrestrial Bacteria—A Rare Event? *FEMS Microbiology Review* 22(2): 79–103.

Obeso, J.R. 2002. The Cost of Reproduction in Plants. *New Phytologist* 155: 321–348.

Richardson, David M., ed. 1998. *Ecology and Biogeography of Pinus.* Cambridge, UK: Cambridge University Press.

Rogers, D.L., and F.T. Ledig, eds. 1996. *The Status of Temperate North American Forest Genetic Resources.* Report No. 16. Davis, CA: Genetic Resources Conservation Program, University of California, Davis.

Sedgley, M., and A. Roderick Griffin. 1989. *Sexual Reproduction of Tree Crops.* New York: Academic Press.

Smith, J.A. 2002. The Precautionary Principle in the Context of an Ecological Paradigm: Some Questions and Values. http://www.earthethics.com/precautionary_principle.htm (accessed November 28, 2002).

Strauss, S.H., W.H. Rottmann, A.M. Brunner, and L.A. Sheppard. 1995. Genetic Engineering of Reproductive Sterility in Forest Trees. *Molecular Breeding* 1: 5–26.

Strauss, S.H., W. Boerjan, J. Cairney, M. Campbell, J. Dean, D. Ellis, L. Jouanin, and B. Sundberg. 1999. IUFRO Position Statement on Biotechnology. http://www.fsl.orst.edu/tgerc/iufro_pos-statm.htm (accessed November 12, 2002).

Sugita, K., E. Matsunaga, and H. Ebinuma. 1999. Effective Selection System for Generating Marker-Free Transgenic Plants Independent of Sexual Crossing. *Plant Cell Reports* 18: 941–947.

Syvanen, M. 1994. Horizontal Gene Transfer: Evidence and Possible Consequences. *Annual Review of Genetics* 28: 237–261.

Thielges, B.A. 1982. A Strategy for Breeding Population-Based Disease Resistance in Forest Trees. In *Proceedings of the Joint IUFRO Meeting of Working Parties in Genetics about Breeding Strategies Including Multiclonal Varieties, Sensenstein, Germany, September 1982.* Staufenberg-Escherode, Federal Republic of Germany: Lower Saxony Forest Research Institute, 194–215.

Van den Belt, H. 2003. Debating the Precautionary Principle: "Guilty until proven innocent" or "Innocent until proven guilty"? *Plant Physiology* 132: 1122–1126.

Walter, C., M.V. Berridge, and D. Tribe. 2001. Genetically Engineered *Klebsiella planticola*: A Threat to Terrestrial Plant Life? www.forestresearch.co.nz/PDF/Ingham-Rebuttal.pdf (accessed November 28, 2002).

Woese, C.R. 2000. Interpreting the Universal Phylogenetic Tree. *Proceedings of the National Academy of Sciences USA* 97(15): 8392–8396.

# 6

# Tree Biotechnology
# in the Twenty-First Century
## *Transforming Trees in the Light of Comparative Genomics*

STEVEN H. STRAUSS AND AMY M. BRUNNER

G enetic engineering (GE) is the physical isolation, modification, and asexual transfer of genes. It is employed when new, qualitative changes in an organism are desired that are difficult to accomplish via traditional breeding. It may involve the creation and introduction of modified forms of native or homologous genes that are rare in wild populations but desirable in tree crop production environments, such as for sterility or semidwarfism. Or it may involve the introduction of exotic genes from distant taxa that impart useful functions, such as the ability to detoxify novel environmental pollutants, or certain pest management traits. It complements, but does not replace, the quantitative, polygenic, and incremental breeding for basic productivity and adaptation to environment of traditional breeding programs.

In contrast to traditional breeding, which selects heritable phenotypes with almost no knowledge of their genetic basis, GE starts from knowledge of molecular physiology of an organism. It therefore is closely tied to advances in basic understanding of gene-physiology-environment-trait relationships, the information for which arises primarily from the study of model organisms and general advances in molecular biology. Most of the knowledge base for GE therefore comes from outside the forestry and arboriculture sectors. This contrasts strongly with traditional breeding, which is largely concerned with management of local or regional gene pools for specific species. GE is an entirely new scientific basis for tree breeding, and it therefore poses severe challenges to the traditional structure of breeding, forest genetics, and commercial forestry. Commercial breeding goals must be clearly articulated if the precision of molecular breeding is to be employed. Traditional breeding often moves gradually toward poorly defined elements of higher productivity, with little understanding of how it is proceeding in a physiological or ecological sense. Industry is also rarely able to specify what wood quality attributes will be advantageous in 10 to 50 years. Molecular biology cannot aid movement toward a place that cannot be specified, and such precision appears to be beyond the realm of most of today's business and technology practices. Thus, a key initial role for genomics and GE is likely to be in simply helping to define what the goals of tree biotechnology should be by providing a rich new array of breeding options (i.e., gene-trait-value relationships) that do not now exist.

The cost of developing the needed genomic information and GE methods—and then successfully negotiating intellectual property, regulatory, and public acceptance hurdles—is substantial, requiring greater collaboration over a longer term among industries, agencies, and academic organizations than has been traditional in biotechnology. Unfortunately, although the need for biotechnology consortia and cooperatives is therefore greater than ever, poor economic performance and consolidation in the forest industry are driving many of the existing research consortia in the United States to extinction, or to insufficient critical mass. Thus, we believe that the outlook for implementation of all the promising science and technology that we discuss below is, frankly, dismal, at least for the near future.

We first discuss tree breeding and its obstacles, then explore how GE—by enabling capture of the rapidly expanding knowledge of plant genomes and molecular physiology—might solve some major problems inherent in tree domestication. We then return to the considerable structural challenges from scientific, social, and business perspectives that stand in the way of technological progress.

## Tree Breeding and Domestication

Wherever humans have chosen to cultivate trees for specific products—whether fruit, wood, pulp, or energy—breeding has played a large part in boosting yields and adapting trees to the new environments and economics of production systems (e.g., Zobel and Talbert 1984). Some of the most dramatic examples come from the use of trees as exotics, where major changes in form, pest resistance, and yield often accompany crop development in novel environments and social contexts (Figure 6-1). The extent of genetic diversity in trees, and thus of the opportunity for genetic change, is most readily seen where trees have been deployed as clonal, vegetative propagules, as in some poplars (genus Populus) and eucalypts (genus Eucalyptus). Such deployment allows the genetic variation present, including the nonadditive forms such as heterosis, to be readily observed and a large proportion of it captured. However, most tree species are not deployed as clonal propagules

**FIGURE 6-1.** Change in Tree Form and Growth during Domestication.

*Pinus radiata* in New Zealand after a single generation of selection for form and productivity (left), compared to trees from wild progenitors in California (right). The consequences of the increased growth and straightness for wood yield and product quality are obviously very large. This rapid change was enabled by the existence of highly polymorphic alleles for these traits. Other desirable changes in form and growth that may have large productivity benefits, but would be opposed by natural selection in the wild (e.g., absence of sexual reproduction), will be extremely difficult to achieve via conventional breeding. GE will be a far more effective means for domestication for these kinds of traits. Photo used with permission: Pauline Newman, Forest Research—New Zealand.

and have undergone very little domestication of any sort. In contrast to most domesticated crops, only specialists would be able to distinguish bred trees from their wild relatives.

Domestication of trees has been limited because serious breeding efforts are relatively recent and because of constraints to breeding inherent in the life history and management characteristics of trees (Bradshaw and Strauss 2001a). The main obstacles to domestication of trees are the following:

- the multiyear delay until the onset of flowering, which slows breeding cycles,
- intolerance of inbreeding that prevents the fixation of desirable alleles and genotypic configurations as inbred lines,
- large size and slow development, which make establishment and measurement of large breeding populations costly,
- low uniformity of the growth and testing environment when species are planted on nonagricultural lands, reducing trait heritability,
- the long time frames that are needed to assess tree health and productivity,
- limited alteration of the production environment toward agronomic conditions, reducing the ability to modify genotypes to adapt to altered, productivity-enhancing conditions, and
- the perennial life cycle and the consequent need for adaptation to unpredictable annual climatic cycles, requiring conservativeness in alteration of adaptive characteristics during domestication.

## Changes in Breeding Enabled by GE

GE cannot change the constraints imposed by the environment, perennial growth, and the need to assess tree health over several years and sites. These are fundamental constraints of the biology of the organism and production system. However, for genes with little effect on organism physiology and adaptation (e.g., herbicide resistance), it should be possible to shorten the selection cycle once the extent of somaclonal variation—the background mutation imparted by the gene transfer process—is understood. We should in the near future attain such understanding for poplars, in which very low levels of somaclonal variation appear to be present and highly stable, productive transgenic lines can be identified in the first few years of testing (e.g., Meilan et al. 2002 and Chapter 4; Figure 6-2). However, for novel transgenic changes that are expected to have multiple pleiotropic effects on productivity and physiology (e.g., significant changes in lignin chemistry), the testing cycle may need to be longer than is currently the case—at least during initial stages—to ensure that unexpected changes in adaptation to abiotic and biotic stresses have not been imparted. As with other forms of novel breeding, the extent of testing needed will have to be determined empirically, via systematic trial and error (commonly referred to as "adaptive management") during early commercial applications.

### Speeding Flowering and Clonal Propagation

GE may be able to accelerate breeding of trees by speeding the onset of sexual reproduction and by facilitating clonal propagation from a diversity of geno-

**FIGURE 6-2.** Example of Healthy Transgenic Trees Growing in a Research Trial in the United States.

The trees in the left row comprise a number of different transgenic versions of a single hybrid cottonwood clone (*Populus deltoids* x *nigra*), and those on the right are nontransgenic versions of the same clone. The great majority of the transgenic trees showed normal growth and stable transgene expression, as have most field trials with transgenic trees. Photo used with permission: R. Meilan and S. Strauss, Oregon State University.

types and maturation states. The Arabidopsis genes *LEAFY* and *APETALA1* can speed flowering and provide normal seed set in citrus trees, potentially accelerating breeding markedly (Peña et al. 2001). *LEAFY* also speeds flowering in poplars but is ineffective in many genotypes and does not appear to result in production of normal pollen or seed production (Rottmann et al. 2000). Nonetheless, these results inspire confidence that ultimately genes and attendant expression control systems will be identified that can serve as genetic switches to allow flowering to be induced precociously in trees (reviewed in Martin-Trillo and Martinez-Zapater 2002; Brunner et al. 2002). Because of the genetic complexity and evolutionary diversity of mechanisms of flowering onset (reviewed in Battey and Tooke 2002), these may need to be different genes in different tree taxa.

Vegetative propagation is the most direct means for making large genetic gains because it allows a large proportion of the extensive genetic diversity in trees to be captured in a single cycle of selection. However, the extent to which genotypes submit to various propagation methods, such as rooting of cuttings, itself shows very high genetic variation, and as most trees age they lose the capacity for the "rejuvenation" required for effective clonal propagation and derivation of healthy, fast-growing clonal "seedlings" (reviewed in Brunner et al. 2002). The capability for rejuvenation and the related trait "cellular regeneration capacity" also tend to show very high genetic variation, and may be controlled by major genes (e.g., Han et al. 1995). This suggests that it is possible to identify genes whose selective induction could promote the efficiency of clonal propagation and perhaps enable lower cost methods to be employed (e.g., by avoiding the need for cryostorage to maintain juvenility). Fundamental studies of the genes that control epigenetic states, such as histone methylation (reviewed in Reyes et al. 2002), may also provide new means for reversing epigenetic programs, and studies of genes that control apomixis (Grimanelli et al. 2001) may someday provide new ways to induce the formation of seedlike clonal propagules. Finally, and most important for GE, methods for enhancing rejuvenability/totipotency of plant cells should enable increased rates of transformation and regeneration of transgenic plants—currently the chief bottleneck to the use of transformation as a breeding and research tool.

## Inserting Novel Alleles

GE derives its greatest novelty from the ability to isolate and selectively modify genes. All other breeding methods work from phenotypes or statistical tools for inferring abstract genetic properties from related individuals. Because of the conservation of the genetic code, genes from other organisms, such as those for herbicide resistance, as well as desirable alleles that are rare in the gene pool or present only as recessive forms, can be effectively used in breeding via conversion to dominant forms and insertion into a variety of genotypes. This is the case with traits such as reproductive sterility. Male-sterile genotypes, though rare, can be readily identified in wild populations of many species; however, they often involve organelle genomes or recessive loss-of-function nuclear genes (e.g., Schnable and Wise 1998). Using methods such as RNA interference (RNAi: e.g., Matzke et al. 2001; Wesley et al.

2001) or promoter:cell toxin fusions (ablation: e.g., Nilsson et al. 1998), dominant loss-of-function alleles can be created anew and inserted (or crossed) into a diversity of genotypes with dominant trait inheritance.

Because plant domestication often involves alterations in plant phenotypes that make them more productive under high-density, high-fertility, monoclonal plantings, it often involves the loss of functions important to promoting fitness in biotically and environmentally diverse wild populations. For example, dwarf varieties are commonly used in grains and highly desirable in tree orchards, and several genes for dwarfism have been identified (Silverstone and Sun 2000). Similar changes may be desirable in forest plantation trees, as short, narrow-crowned trees may be more productive in monoclonal wood plantations if the population invests less in roots, branches, and reproductive tissues relative to stems, and more stems can be packed per unit area (Bradshaw and Strauss 2001a). Dwarfing genes, because they would reduce the fitness of wild forest trees dramatically, may also provide a large biosafety benefit by making transgenic progeny uncompetitive. Alleles that cause such deleterious changes in form would be very difficult to find in wild populations but should be readily produced via GE.

Likewise, genes that change wood structural or chemical qualities to improve its value as an industrial feedstock will tend to move trees away from the normal range produced by natural selection. It will therefore be difficult to identify many such alleles in the wild, and thus very difficult to increase their frequency via traditional population breeding in trees. However, they also could be readily produced via GE (Campbell et al. 2003). Genes that affect lignin quantity and chemistry are examples; they could be of tremendous economic and ecological value by reducing the great cost, and byproducts, of lignin removal during pulping (Sedjo, Chapter 3) and might possibly also increase fiber productivity and quality (Hu et al. 1999; Dinus et al. 2001).

# Translating Genomic Knowledge into Tree Biotechnology

## Comparative Genomics

In addition to the ability to produce and mobilize alleles for desired traits that would not be accessible to traditional breeding, another major advantage of GE is that it allows us to capitalize on knowledge from comparative genomics and molecular biology (Figure 6-3). This is enabled by the extensive conservation of gene and derived protein sequences, the availability of powerful computer methods for searching and comparing large sequence databases, and the extensive functional and microsyntenic (map position) conservation within broad taxonomic classes of plants. Thus, large-scale

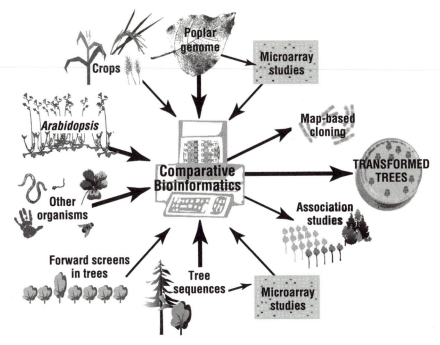

**FIGURE 6-3.** Genomics Information Important to Molecular Domestication of Forest Trees.

Bold items and arrows show information and analysis expected to be of most importance to genomic analysis or commercial deployment. Because of the high degree of conservation of the genetic content of organisms, especially within plants, genomic information about any species can be used to suggest routes for study, marker-aided breeding, or transgenic modification of trees. Information from intensively studied species, such as *Arabidopsis*, and fundamental studies of molecular genetic biology in other model organisms are expected to be most important because they will contain a large amount of information on gene function as well as sequence. This information, plus functional genomic studies from other crop and plant species, can be used to identify candidate genes in databases of tree genomes that may be useful for modifying economically important traits. Poplar will be of particular value because of its soon-to-be-completed genome sequence and its amenability to transformation—the latter allowing hypotheses of gene function and trait modification in trees to be directly and rigorously tested. Forward screens in trees, where transformation is used to create tagged mutants, is also expected to be useful for species such as poplar, where transformation can be conducted on a large scale. High-throughput studies of gene expression through microarray hybridization allow new genes from trees to be recognized based on expression patterns and the expression characteristics of large numbers of homologous genes to be verified. Recognition of gene homologies with other organisms via sequence or synteny will aid identification of genes in map-based cloning experiments and in selection of genes for studying associations of natural variants in traits with sequence polymorphisms. Because of the costs and statistical difficulties of fine mapping for quantitative trait loci (QTL) in most tree species, transformed trees will be needed for conclusively demonstrating the functional roles of suspected allelic variants for many important traits, as well as for deploying dominant alleles that can impart domestication phenotypes in the variety of genetic backgrounds needed for commercial applications.

mutagenesis projects in model organisms (e.g., the *Arabidopsis* 2010 project: Chory et al. 2000) should produce a great deal of information useful to molecular breeders of any plant species. By contrast, traditional breeding is necessarily confined to knowledge from the species or closely related taxa.

For example, for genes that affect wood chemistry it will often be profitable to look to studies of model organisms—particularly where there are complete genomes, extensive EST sequence banks, or large-scale expression/function assays—to try to identify key transcription factors that efficiently control wood quality in trees. Likewise, because of the complexity of flowering it will nearly always be preferable to study candidate genes identified by extensive studies of flowering genes in rapid-cycling model organisms, rather than to try to isolate genes for this process de novo from trees. However, large sequence databases, either from genome or expressed genes (cDNA), must be available in trees for comparative genomic methods to proceed. Therefore, one of the great benefits of structural genomics projects in trees is the *in silico* "functional" access that structural genomics provides to studies in model organisms. Alternatively, though far less powerful, if there is sufficient gene conservation—and the orthologous gene family members can be targeted (no small task for most genes)—specific genes can be chosen for isolation in tree species based on their sequence and putative functional conservation.

Even where large sequence banks exist for tree species, to select the best genes from among the numerous (often hundreds) of candidate genes and gene families for control of specific traits is challenging. Gene deletion, duplication, mutation, rearrangement, and functional divergence often make recognition of functional evolutionary orthologs difficult. The same gene or its functional ortholog may perform differently depending on its organismal context. Moreover, evolutionary changes in key genes can alter entire regulatory networks. For example, changes in the level of an encoded protein due to gene duplication/deletion or novel interactions among proteins can modify a signaling pathway as well as the extent of cross-talk and redundancy among pathways. Thus, some kind of functional assay in trees is usually required. The three basic experimental options are association studies, microarray expression analysis, and production of transgenic trees.

**Association Studies.** This method searches for statistical associations between traits and natural genetic variants of candidate genes in wild or experimental populations. Association studies will help to define effective targets for GE, as well as provide options for marker-aided breeding. However, there are a number of serious obstacles to their application. The main constraints are the limited knowledge of potential candidate genes underlying specific traits and the difficulty, statistical risk of Type I errors, and cost of screening very large numbers of candidate loci and progeny. Large numbers need to be screened because of low linkage disequilibrium in outbred tree populations and high background environmental and genetic variability. Phenotyping of progeny can also be a major cost, especially for complex traits such as wood chemistry, or delayed-expression traits such as onset of flowering.

**Array Expression Studies.** Scanning large numbers of genes from trees for patterns of expression offers another experimental tool. Array expression studies require a large sequence bank of genes, derived via either cDNA or genomic sequencing, and thus they are also a route for novel gene discovery (reviewed in Aharoni and Vorst 2002). Key constraints are the costs of generating the sequence data and the fact that even large EST banks usually fail to include a large proportion of genes. EST databases are usually biased toward highly expressed "structural" genes, whereas weakly expressed transcription and signal transduction factors can play key roles in generating phenotypic variation (Doebley and Lukens 1998). Conducting and statistically analyzing large arrays from a diversity of experiments are also problematic (e.g., Quackenbush 2001) and generally require verification studies with more precise, single gene or small-array methods. A key advantage of array studies is that the method allows many genes, including many "novel" genes (i.e., ones without clear homologs/orthologs in other species), to be identified. It is therefore also a method of de novo gene discovery. By allowing global expression changes to be monitored under different environmental conditions, at different time points following an environmental stimulus, at different ages and stages of development, and in different tissues, microarray studies can point to the most appropriate gene(s) to manipulate for a desired phenotype. For example, such studies may distinguish genes that are likely to have pleiotropic effects or undesired additional effects under a particular environmental stress from a gene that acts downstream in a single biochemical pathway (e.g., Chen et al. 2002). An in-depth microarray study of xylem differentiation in poplar successfully clustered large numbers of known and novel genes (Hertzberg et al. 2001).

**Transgenic Studies.** Finally, researchers can produce directed mutants—via impairment/loss of gene expression or hyperexpression—in an isogenic background. Transformation provides the most informative and precise information about the basic physiological function of genes and thus about what might be accomplished with them via GE. The main constraint is that high-throughput GE is not possible in most tree species, with the notable exception of some poplar genotypes. Even here, however, the cost of generating the tens to hundreds of transgenics desirable for investigating the roles of a number of candidate genes for a specific trait is considerable. And because of the variation in transgene expression with each gene transfer event, it is necessary to generate and both phenotypically and molecularly evaluate a number of independent lines per gene. Finally, because of the strong canalization and redundancy of developmental mechanisms—and the presence of many gene family members with similar roles—single gene knockouts often fail to display obvious phenotypes. Seeing the effects of specific genes therefore often requires multigene suppression or gain-of-function transgenes, such as those with ectopic overexpression. However, gain-of-function mutations are much harder to interpret in terms of native function. Inducible or tissue-specific overexpression is preferable, but is rarely employed, for large-scale screens. The recent demonstration (Abbott et al. 2002) that three unrelated endogenes were coordinately suppressed by a single chimeric transgene suggests that multigene suppression can become a feasible approach. Adapting this approach to large-scale studies will likely require use of the more efficient RNAi method (rather than cosuppression) for inducing gene silencing (Wesley et al. 2001) and development of a high-throughput system for constructing vectors. For example, Gateway technology (Invitrogen Corporation), which is based on site-specific recombination, could be adapted for this use (e.g., Wesley et al. 2001).

The main obstacle to comparative genomics is therefore the cost of conducting the large-scale sequencing and/or functional studies needed. As with other organisms, this problem is best addressed by large, multi-investigator, internationally coordinated efforts to establish basic sequence/informatics databases and national or international centers for efficient functional assays. Unfortunately, private databases, such as the large EST databases for *Pinus* and *Eucalyptus* produced by Genesis/Arborgen, are generally not available to others, and public efforts, although somewhat coordinated, have often been duplicative (e.g., multiple studies of poplar cambial tissues). Also the international community has received only limited printed microarrays from expression centers, impeding standardization of array experiments. However, the upcoming sequencing of the poplar genome by the U.S. Department of Energy has been well planned internationally, and its informatic analyses and resulting database composition are expected to benefit from provision of data from several countries (International Populus Genome Consortium: http://www.ornl.gov/ipgc). In addition, 60,000 *Pinus taeda* ESTs from wood-forming tissues are freely available, and new U.S. National Science Foundation–funded projects will be adding over 100,000 ESTs from additional tissues to this collection.

## Novel Gene Discovery in Trees

Because of the different phylogenetic and adaptive histories of trees, there are likely to be many genes that control important "tree-specific" traits that will be missed if only genome knowledge from model organisms is used to select

genes for investigation. It is therefore highly desirable to independently identify developmental regulatory genes in trees. There are three means of gene
discovery in trees: map-based cloning, large-scale DNA sequencing, and gene
tagging.

**Map-Based Cloning.** Use of genetic linkage mapping of natural genetic variants can identify the underlying genes. Map-based cloning depends on the
availability of natural or synthesized populations that have substantial linkage
disequilibrium (such as a single full-sib family), and that has segregating variation in genes that affect a target trait. Wild populations could also be
employed, but that would require a density of markers that is currently
unavailable and unaffordable for any tree species. For example, in landraces of
maize, disequilibrium declines rapidly beyond 200 bp (Tenaillon et al. 2001);
a similar level of disequilibrium in trees would therefore require hundreds of
thousands of markers to identify unknown genes. In pedigree populations,
the most significant obstacles are the need for very large populations to be
screened to enable sufficient genetic precision for physical isolation and the
difficulty of fine-mapping with quantitative trait loci (QTL; Flint and Mott
2001). For example, even a highly heritable trait such as vegetative growth
phenology appears when studied intensively in specific environments and
clonally replicated genotypes to be controlled by numerous genes of small
effect (Jermstad et al. 2001). More complex and economically important traits
such as yield or wood quality are likely to prove far more intractable. This will
make QTL localization extremely difficult. Moreover, in most cases it would
still be necessary to transform with a number of candidate genes from a
mapped region to identify the causative one, assuming that its effects are at
least partially dominant and statistically detectable (Mackay 2001). Map-
based cloning therefore appears to be technically infeasible for the large
majority of tree species and genes in the foreseeable future. However, for tree
genera in which EST collections are available and large SNP surveys are feasible, the use of EST polymorphisms should provide new hypotheses about the
genes that cause trait polymorphism (Brown et al. 2001; Cato et al. 2001).

**Identifying Novel Genes through Genome and EST Sequencing.** Either array hybridization or high-throughput transformation can be used to determine the effects of novel (or highly dissimilar) genes that have been revealed
in sequencing projects. For example, genes important to onset of flowering
might be identified by their associated changes in expression with tree age/
reproduction (cf. Chen et al. 2002). Gene selection could be aided by searching for motifs such as DNA binding or kinase domains that give indications
of broad function, and transformation studies could be conducted after
informatic and array studies suggest which genes are most likely to affect tar-

get traits and processes. The costs and complications of these methods are similar to those discussed previously for array and transformation methods in comparative genomics.

**Transformation-Based Gene Tagging.** Random gene insertion can be used as a means to identify novel genes based on developmental phenotypes or reporter-gene expression. The vast majority of genes that have been identified in genetics have been found through the analysis of mutants. Mutations are generally caused by recessive, loss-of-function alleles that are only revealed through inbreeding and consequent homozygosity. Despite the vast abundance of natural mutations in tree populations, because of long generation times, intolerance of inbreeding, and poor genetic maps, large mutant populations have neither been created nor screened for the causative genes. However, two recent techniques for creating tagged, dominant mutants in trees facilitate direct mutation-based gene isolation.

- *Activation tagging.* Activation tagging is a method whereby a strong enhancer of gene expression—active up to several kb from a promoter—is randomly inserted into the genome, disturbing (elevating) expression of a nearby gene (Weigel et al. 2000). This gain of function causes dominant phenotypes, and the presence of the known transgene sequence—generally with an antibiotic resistance marker to facilitate plasmid rescue—enables the affected portion of the genome to be isolated and sequenced. Ultimately, a candidate gene nearest to the enhancer is identified and then retransformed to verify that it is the cause of the phenotype. This method yields tagged phenotypes in about 1% of independent transgenic lines, including in poplars (Ma et al. 2001). It therefore requires that very large populations of transgenics and rapid screening methods for specific traits be available. Moreover, the complex rearrangements that can occur during transformation require that putatively tagged genes be retransformed to verify their association with phenotypes (cf. Tax and Vernon 2001). A large-scale tagging project in a tree—even poplar—is therefore not a small task.
- *Gene/promoter trapping.* Gene/promoter trapping is a method whereby a reporter gene, such as GUS, is randomly inserted into the genome either with a basal promoter (allowing it to be expressed whenever it lands by an enhancer) or without, requiring direct incorporation into a gene for expression (reviewed in Springer 2000). The population of transgenics is then screened for developmental-, tissue-, cell-, or environment-induced expression patterns of interest. This method yields a considerably higher number of tagged mutants than does activation tagging and rapidly identifies promoters that may have desired expression properties, but does not itself give developmental mutations. A large number of vascular expressed genes have been identified in this manner in poplar (A. Groover, USDA Forest Service, personal communication).

# Refining Experimental Transgenics into Commercial Products

## Creating New Alleles for Target Traits

If the social controversies over GE do not curtail funding for research, the next decade or two could be productively spent both in scanning genomes using comparative and de novo gene discovery methods and in testing proof-of-concept transgenic trees for the ability to modify important traits. The traits will certainly include mainstays such as yield, wood structure, and wood chemistry but could also include more novel and specific traits such as crown structure, flowering, vegetative propagation capacity, transformability, growth rhythms, and biotic and abiotic stress resistances. The complete genome sequence of poplar—the model forest tree—will soon be available, and tackling it vigorously in an internationally coordinated manner, via both array and transgenic studies, would be highly desirable. It would also be desirable to conduct similar studies in taxonomically distinct forestry species, such as conifers and eucalypts. However, studies in those species would be likely to rely less on array studies than on functionally more informative transformation tests. In a decade it might be possible to have identified a large number of new genes that are likely to, or have been demonstrated in transgenics to, affect important traits in trees. However, much additional work would then be required to develop actual commercial varieties.

In many cases simple gene knock-outs or overexpression phenotypes will be inadequate for delivering the precise phenotypic changes desired in a domesticated tree. For example, excessive lignin reduction is likely to be deleterious, but an intermediate level may be both ecologically safe and economically valuable. It will therefore be necessary to identify the best genes of several that can affect a given trait and then to learn how to engineer the best alleles for those genes (Figure 6-4). RNAi may be adequate where reductions in gene activity are required, but too little is known at present about its stability and quantitative consistency during development and environmental variation. Alternative means for modulating gene expression, such as by using designed zinc-finger transcription factors for transcriptional repression or activation (Beerli and Barbas 2002), may be needed for more refined control of gene expression. In addition, facile methods for coordinate suppression or overexpression of multiple genes, such as use of a single polyprotein transgene to express multiple proteins, will be needed to generate many desired phenotypes (reviewed in Halpin et al. 2001).

Likewise, there are too few native promoters and reliable means for inducing, or elevating, gene expression in tree species. This will be important where high levels of expression of foreign genes are desirable in specific tis-

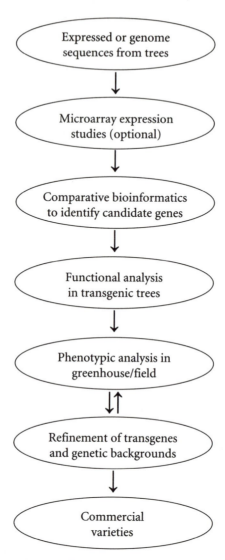

**FIGURE 6-4.** Summary of Steps in Tree Domestication via Comparative Bioinformatics and Transformation.

sues to enhance activity while avoiding unintended ecological effects (e.g., for production of industrial or bioremediation proteins). Thus, if we are to be successful in generating genomic information that leads to useful transgenic varieties, it will be important not only to conduct large screens of gene function but also to learn to engineer the most promising genes so that their effects can be carefully controlled. Toward this goal, system-level genomic tools, such as arrays, that allow the monitoring of changes in metabolism imparted by transgenes, may be useful for detecting adverse pleiotropic effects as early as possible. For example, microarray data might be helpful in verifying that genetic circuitries for responding to herbivory are not unduly impaired by lignin modification genes. The analysis of array data to infer the function, and the responses to perturbations, of complex regulatory networks is currently the subject of intensive bioinformatics research (e.g., Gifford 2001; Wyrick and Young 2002).

## Ecological and Social Issues

Despite the abundant genetic diversity present in forest tree species, because of their outcrossing system of mating, long generation times, very large effective population sizes, strong constraints from natural selection, and limited efforts at domestication, there have been only modest changes to tree form, productivity, or wood properties. Even the poplars and eucalypts, where marked improvements of productivity have resulted from interspecific hybridization combined with clonal propagation, have still retained their basic form and properties. The tools of genetic engineering can clearly move species beyond all prior breeding technologies and into new developmental zones. For that to occur, however, requires a measure of caution and skepticism, as well as technological enthusiasm for the possibilities (see more extensive discussions of risk by Burdon and Walter, Chapter 5).

Although we know that one or a few major genes can produce domestication traits in crops (e.g., Paterson et al. 1995; Bradshaw and Strauss 2001a), it is unclear whether adapted varieties can be produced with those major gene changes if the many polygenic modifications that also accompanied annual crop breeding during domestication are not also present. In forestry, such fine-tuning can be done to a much more limited extent, mainly by careful selection of transgenic events and genetic backgrounds, and by extensive field testing prior to widespread use. The need to assess a diversity of transgenic lines and genetic backgrounds further underlines the importance of gene transfer efficiency. If there is a single kind of trait that is most important to molecular domestication via transformation—and which therefore should be screened for vigorously in functional genomic screens—it may be the capacity for induced regeneration of plants from single cells via either embryogenic or organogenic pathways. Although great advances have been made in transformation methods to use positive phenotypic markers of selection and to enable the elimination of unwanted transgenes such as those for antibiotic resistance (e.g., Ebinuma et al. 2001), there are still no transgenes whose expression can increase transformation/regeneration frequency substantially in a variety of genotypes and species. However, recent work has identified a number of promising candidates (Banno et al. 2001; Tantikanjana et al. 2001; Boutilier et al. 2002; Prakash and Kumar 2002).

The main risks of using novel, highly domesticated trees will be to the growers and local economies, rather than to wild ecosystems, as highly altered trees are unlikely to be competitive in wild environments. However, the public may insist on a high degree of precaution until a record of safety is established. The novelty of such trees may also force a social requirement that all feasible steps be taken to confine them to plantations, requiring some form of engineered infertility (Peña and Séguin 2001). Thus, in addition to genomic scans it will be important to continue developing and testing the efficiency of infertility transgenes. The best means for doing this might be through release of some commercial horticultural products whose only trait is male and/or female sterility (to reduce allergens and nuisance tissues), combined with monitoring of gene flow or other effects (cf. NRC 2002). This would allow long-term, commercial-scale "testing," under highly diverse environments, of the reliability of engineered sterility with little or no eco-logical consequence should it fail. Rigorous testing of infertility systems is also likely to be required before genes that might increase fitness in wild pop-ulations, such as novel pest toxin genes like those from *Bacillus thuringiensis*, could be deployed (e.g., Raffa, Chapter 13).

The scientific and technological possibilities for forest biotechnology are vast. However, the costs, social concerns, and ecological unknowns are also considerable (Doering, Chapter 8). It seems clear that progress will require collaboration on many fronts, as no one organization—academic, govern-ment, corporate, or consortium—can do all that is needed. In this respect we suggest that broad patents on gene sequences and functions—unless clearly tied to utility in specific commercial tree taxa—be avoided whenever possible (Williamson 2001). The prospect of large license costs and legal fees to negoti-ate the intellectual property maze, combined with the already very consider-able regulatory and public relations costs of biotechnology, clearly discourages the broad interest and support from the forest industry, small landowners, and the professional natural resource community that are needed for progress to occur. The associated secrecy also fosters the public's fears about genetically modified organisms (cf. Bobrow and Thomas 2001), as well as concerns of potential liability for genetic pollution from a company's patented genes. We encourage companies and researchers to consider carefully the costs of a highly patent- and secrecy-oriented approach to research. We believe that the presumed benefits that such an approach may provide to single companies in the short term may be overshadowed by the enormous, cumulative burden it places on the entire forest biotechnology enterprise.

## Conclusion

Because of the conservation of the genetic code—and of basic developmental programs within plants—genes can be isolated, modified, and reinserted into the same species, or genetic information can be moved across species, to produce effective trait modifications. As a new method for improving the productivity of planted forests, GE could be important to meeting the challenge of preserving biodiversity through protection of wild forests from exploitation; sequestering carbon; replacing fossil energy and materials with renewably generated biomaterials; providing new means for bioremediation of polluted lands and water; and producing trees that are better able to cope with a variety of biotic and abiotic anthropogenic stresses (Fenning and Gershenzon 2002).

When altering developmental properties, it will usually be most effective to use knowledge gained from comparative genomic studies to identify and modify native genes. However, when entirely new traits are desired, such as the production of new chemicals (Rishi et al. 2001) or new bioremediation capabilities (Merkle and Dean 2001), exotic genes will be required, generally in combination with some kind of genetic infertility system to control spread into wild environments. In the near term, GE will require a clonal propagation system for delivering modified trees to plantations (Griffin 1996), itself a major constraint to economical use of GE in most forestry systems. However, genes that enable flowering on demand and increase clonal propagation efficiency may result in transgenes being deployed widely in breeding programs and sexual propagules in the future.

GE and gene transfer enable the production and wide deployment of novel, dominant "domestication" alleles that may have large benefits for productivity or product quality under intensive plantation management systems (Bradshaw and Strauss 2001b). These genes, which reduce the competitiveness and reproductive success of single trees, are virtually inaccessible to traditional breeding because of their adverse effects on fitness in wild populations and consequent rarity and recessive gene action. They should therefore pose extremely little threat of "invasion" or disruption of intact wild populations via introgression.

The main technical obstacles to GE are the difficulty and costs of transformation and clonal propagation for most commercially important species and the very limited knowledge of structural and functional comparative genomics in trees. Biosocial constraints include the lack of effective demonstration of a genetic confinement system, which may be needed to convince the public of the environmental safety and legality of transgenic plantations, and strong social resistance to GE in forestry by some sectors of the public (Strauss et al. 2001). As widely agreed at a recent international conference (Bradshaw and Strauss 2001a), research is the most important need for development of forest biotechnology, particularly (a) work on improvements in the efficiency of transformation, cellular regeneration, and clonal propagation to speed research and testing; (b) the expansion of functional genomic knowledge, so that the identities of genes that can modify economically important traits are known; and (c) field demonstrations of the value and biosafety of genetically engineered trees.

A serious research commitment must be made soon if GE trees are to be available for significant commercial uses in one to two decades. The scientific possibilities are vast. However, because of the reductions in research support by industry, the very limited public funding of applied forest biotechnology, and the high level of public concern over crop GE in some countries, the prospects for large near-term applications—at least in the more developed world—appear limited.

# References

Abbott, J.C., A. Barrakate, G. Pincon, M. Legrand, C. Lapierre, I. Mila, W. Schuch, and C. Halpin. 2002. Simultaneous Suppression of Multiple Genes by Single Transgenes: Down-Regulation of Three Unrelated Lignin Biosynthetic Genes in Tobacco. *Plant Physiology* 128: 844–853.

Aharoni, A., and O. Vorst. 2002. DNA Microarrays for Functional Plant Genomics. *Plant Molecular Biology* 48: 99–118.

Banno, H., Y. Ikeda, Q.-W. Niu, and N.-H. Chua. 2001. Overexpression of *Arabidopsis ESR1* Induces Initiation of Shoot Regeneration. *Plant Cell* 13: 2609–2618.

Battey, N.H., and F. Tooke. 2002. Molecular Control and Variation in the Floral Transition. *Current Opinions Plant Biology* 5: 62–68.

Beerli, R.R., and C.F. Barbas III. 2002. Engineering Polydactyl Zinc-Finger Transcription Factors. *Nature Biotechnology* 20: 135–141.

Bobrow, M., and S. Thomas. 2001. Patents in a Genetic Age. *Nature* 409: 763–764.

Boutilier, K., R. Offringa, V.K. Sharma, H. Kieft, T. Ouellet, L. Zhang, J. Hattori, C.-M. Liu, A.A.M. van Lammeren, B.L.A. Miki, J.B.M. Custers, and M.M. van Lookeren Campagne. 2002. Ectopic Expression of BABY BOOM Triggers a Conversion from Vegetative to Embryonic Growth. *Plant Cell* 14: 1737–1749.

Bradshaw, H.D., Jr., and S.H. Strauss. 2001a. Breeding Strategies for the 21st Century: Domestication of Poplar. In *Poplar Culture in North America,* edited by D.I. Dickmann, J.G.T. Isebrands, J.E. Eckenwalder, and J. Richardson. Ottawa, Canada: NRC Research Press, National Research Council of Canada, 383–394.

———. 2001b. Plotting a Course for GM Forestry. *Nature Biotechnology* 19: 1104

Brown, G.R., E.E. Kadel III, D.L. Bassoni, K.L. Kiehne, B. Temesgen, J.P. van Buijtenen, M.M. Sewell, K.A. Marshall, and D.B. Neale. 2001. Anchored Reference Loci in Loblolly Pine (*Pinus taeda* L.) for Integrating Pine Genomics. *Genetics* 159: 799–809.

Brunner, A.M., B. Goldfarb, V. Busov, and S.H. Strauss. 2002. Controlling Maturation and Flowering for Forest Tree Domestication. In *Transgenic Plants: Current Innovations and Future Trends,* edited by C.N. Stewart. Wymondham, UK: Horizon Scientific Press.

Campbell, M.M., A.M. Brunner, H.M. Jones, and S. H. Strauss. 2003. Forestry's Fertile Crescent: The Application of Biotechnology to Forest Trees. *Plant Biotechology Journal.*

Cato, S.A., R.C. Gardner, J. Kent, and T.E. Richardson. 2001. A Rapid PCR-Based Method for Genetically Mapping ESTs. *Theoretical and Applied Genetics* 102: 296–306.

Chen, W., N.J. Provart, J. Glazebrook, F. Katagiri, H.-S. Chang, T. Eulgem, F. Mauch, S. Luan, G. Zou, and S.A. Whitham. 2002. Expression Profile Matrix of *Arabidopsis* Transcription Factor Genes Suggests Their Putative Functions in Response to Environmental Stresses. *Plant Cell* 14: 559–574.

Chory, J., J.R. Ecker, S. Briggs, M. Caboche, G.M. Coruzzi, D. Cook, J. Dangl, S. Grant, M.L. Guerinot, A. Henikoff, R. Martienssen, K. Okada, N.V. Raikhel, C.R. Somerville, and D. Weigel. 2000. National Science Foundation-Sponsored Workshop Report: The 2010 Project. *Plant Physiology* 123: 423–425.

Dinus, R.J., P. Payne, M.M. Sewell, V.L. Chiang, and G.A. Tuskan. 2001. Genetic Modification of Short Rotation Poplar Wood Properties for Energy and Fiber Production. *Critical Reviews Plant Science* 20: 51–69.

Doebley, J., and L. Lukens. 1998. Transcriptional Regulators and the Evolution of Plant Form. *Plant Cell* 10: 1075–1082.

Ebinuma, H., K. Sugita, E. Matsunaga, S. Endo, K. Yamada, and A. Komamine. 2001. Systems for the Removal of a Selection Marker and Their Combination with a Positive Marker. *Plant Cell Reports* 20: 383–392.

Fenning, T.M., and J. Gershenzon. 2002. Where Will the Wood Come From? Plantation Forests and the Role of Biotechnology. *Trends in Biotechnology* 20:291-296.

Flint, J., and R. Mott. 2001. Finding the Molecular Basis of Quantitative Traits: Successes and Pitfalls. *Nature Reviews: Genetics* 2: 437–445.

Gifford, D.K. 2001. Blazing Pathways through Genetic Mountains. *Science* 293: 2049–2051.

Griffin, R. 1996. Genetically Modified Trees: The Plantations of the Future or an Expensive Distraction? *Commonwealth Forestry Review* 75: 169–175.

Grimanelli, D., O. Leblanc, E. Perotti, and U. Grossniklaus. 2001. Developmental Genetics of Gametophytic Apomixis. *Trends Genetics* 17: 597–604.

Halpin, C., A. Barakate, B.M. Askari, J.C. Abbot, and M.D. Ryan. 2001. Enabling Technologies for Manipulating Multiple Genes on Complex Pathways. *Plant Molecular Biology* 47: 295–310.

Han, K.-H., H.D. Bradshaw Jr., and M.P. Gordon. 1995. Adventitious Root and Shoot Regeneration *in Vitro* Is under Major Gene Control in an $F_2$ Family of Hybrid Poplar (*Populus trichocarpa* x *P. deltoides*). *Forest Genetics* 1: 139–146.

Hertzberg, M., H. Aspeborg, J. Schrader, A. Andersson, R. Erlandsson, K. Blomqvist, R. Bhalerao, M. Uhlen, T. Teeri, J. Lundeberg, B. Sundberg, P. Nilsson, and G. Sandberg. 2001. A Transcriptional Roadmap to Wood Formation. *Proceedings of the National Academy of Science USA* 98: 14732–14737.

Hu, W.-J., S.A. Harding, J. Lung, J.L. Popko, J. Ralph, D.D. Stokke, C.-J. Tsai, and V.L. Chiang. 1999. Repression of Lignin Biosynthesis Promotes Cellulose Accumulation and Growth in Transgenic Trees. *Nature Biotechnology* 17: 808–812.

Jermstad, K.D., D.L. Bassoni, K.S. Jech, N.C. Wheeler, and D.B. Neale. 2001. Mapping of Quantitative Trait Loci Controlling Adaptive Traits in Coastal Douglas-fir. I. Timing of Vegetative Bud Flush. *Theoretical Applied Genetics* 102: 1142–1151.

Ma, C., R. Meilan, A.M. Brunner, J.A. Carson, J. Li, and S.H. Strauss. 2001. Activation Tagging in Poplar: Frequency of Morphological Mutants. Poster presented at Tree

Biotechnology in the New Millennium, July 2001, Stevenson, WA. http://www.fsl.orst.edu/tgerc/iufro2001/poster_abstracts.pdf (accessed April 4, 2002).

Mackay, T.F.C. 2001. The Genetic Architecture of Quantitative Traits. *Annual Reviews Genetics* 35: 303–339.

Martin-Trillo, M., and J.M. Martinez-Zapater. 2002. Growing up Fast: Manipulating the Generation Time of Trees. *Current Opinion in Biotechnology* 13: 151–5.

Matzke, M.A., A.J.M. Matzke, G.J. Pruss, and V.B. Vance. 2001. RNA-Based Silencing Strategies in Plants. *Current Opinion in Genetics and Development* 11: 221–227.

Meilan, R., D.J. Auerbach, C. Ma, S.P. DiFazio, and S.H. Strauss. 2002. Stability of Herbicide Resistance and *GUS* Expression in Transgenic Hybrid Poplars (*Populus* sp.) during Several Years of Field Trials and Vegetative Propagation. *HortScience* 37(2): 277–280.

Merkle, S.A., and J.F.D. Dean. 2001. Forest Tree Biotechnology. *Current Opinion Biotechnology* 11: 298–302.

Nilsson, O., E. Wu, D.S. Wolfe, and D. Weigel. 1998. Genetic Ablation of Flowers in Transgenic *Arabidopsis*. *Plant Journal* 15(6): 799–804.

NRC (National Research Council). 2002. *Environmental Effects of Transgenic Plants: The Scope and Adequacy of Regulation*. Washington, DC: National Academy Press.

Paterson, A.H., Y.-R. Lin, Z. Li, K.F. Schertz, J.F. Doebley, S.R.M. Pinson, S.-C. Liu, J.W. Stansel, and J.E. Irvine. 1995. Convergent Domestication of Cereal Crops by Independent Mutations at Corresponding Genetic Loci. *Science* 269: 1714–1717.

Peña, A., and A. Séguin. 2001. Recent Advances in the Genetic Transformation of Trees. *Trends in Biotechnology* 19: 500–506.

Peña, L., M. Martin-Trillo, J. Juarez, J. Pina, L. Navarro, and J.M. Martinez-Zapater. 2001. Constitutive Expression of *Arabidopsis LEAFY* or *APETALA1* Genes in Citrus Reduces Their Generation Time. *Nature Biotechnology* 19: 263–267.

Prakash, A.P., and P.P. Kumar. 2002. *PkMADS1* Is a Novel MADS Box Gene Regulating Adventitious Shoot Induction and Vegetative Shoot Development in *Paulownia kawakamii*. *Plant Journal* 29: 141–151.

Quackenbush, J. 2001. Computational Analysis of Microarray Data. *Nature Reviews: Genetics* 2: 418–427.

Reyes, J.C., L. Hennig, and W. Gruissem. 2002. Chromatin-Remodeling and Memory Factors. New Regulators of Plant Development. *Plant Physiology* 130: 1090–1101.

Rishi, A.S., N.D. Nelson, and A. Goyal. 2001. Molecular Farming in Plants: A Current Perspective. *Journal of Biochemistry and Biotechnology* 10: 1–12.

Rottmann, W.H., R. Meilan, L.A. Sheppard, A.M. Brunner, J.S. Skinner, C. Ma, S. Cheng, L. Jouanin, G. Pilate, and S.H. Strauss. 2000. Diverse Effects of Overexpression of *LEAFY* and *PTLF*, a Poplar (*Populus*) Homolog of *LEAFY/FLORICAULA*, in Transgenic Poplar and *Arabidopsis*. *Plant Journal* 22: 235–246.

Schnable, P., and R.P. Wise. 1998. The Molecular Basis of Cytoplasmic Male Sterility and Fertility Restoration. *Trends Plant Science* 3: 175–180.

Silverstone, A.L., and T.-P. Sun. 2000. Gibberellins and the Green Revolution. *Trends Plant Science* 5(1): 1–2.

Springer, P.S. 2000. Gene Traps: Tools for Plant Development and Genomics. *Plant Cell* 12: 1007–1020.

Strauss, S.H., M.M. Campbell, S.N. Pryor, P. Coventry, and J. Burley. 2001. Plantation Certification and Genetic Engineering Research: Banning Research Is Counterproductive. *Journal of Forestry* 99(12): 4–7.

Tantikanjana, T., J.W.H. Yong, D.S. Letham, M. Griffith, M. Hussain, K. Ljung, G. Sandberg, and V. Sundaresen. 2001. Control of Axillary Bud Initiation and Shoot Architecture in *Arabidopsis* through the SUPERSHOOT Gene. *Genes and Development.* 15: 1577–1588.

Tax, F.E., and D.M. Vernon. 2001. T-DNA-Associated Duplication/Translocations in *Arabidopsis*. Implications for Mutant Analysis and Functional Genomics. *Plant Physiology* 126: 1527–1538.

Tenaillon, M.I., M.C. Sawkins, A.D. Long, R.L. Gaut, J.F. Doebley, and B.S. Gaut. 2001. Patterns of DNA Sequence Polymorphism along Chromosome 1 of Maize (*Zea mays* ssp. *Mays* L.). *Proceedings of the National Academy of Science* 98: 9161–9166.

Weigel, D., J.H. Ahn, A. Blazquez, J.O. Borevitz, S.K. Christensen, C. Fankhauser, C. Ferrandiz, I. Kardailsky, E.J. Malancharuvil, and M.M. Neff. 2000. Activation Tagging in *Arabidopsis. Plant Physiology* 122: 1003–1013.

Wesley, S.V., C.A. Helliwell, N.A. Smith, M. Wang, D.T. Rouse, Q. Liu, P.S. Gooding, S.P. Singh, D. Abbott, P.A. Stoutjesdijk, S.P. Robinson, A.P. Gleave, A.G. Green, and P.M. Waterhouse. 2001. Construct Design for Efficient, Effective and High-Throughput Gene Silencing in Plants. *Plant Journal* 27: 581–590.

Williamson, A.R. 2001. Gene Patents: Socially Acceptable Monopolies or an Unnecessary Hindrance to Research? *Trends in Genetics* 17: 670–673.

Wyrick, J.J., and R.A. Young. 2002. Deciphering Gene Expression Regulatory Networks. *Current Opinion in Genetics and Development* 12: 130–136.

Zobel, B., and J. Talbert. 1984. *Applied Forest Tree Improvement.* New York: John Wiley.

# PART II

# *Ethical, Social, and Ecological Caveats*

# 7

# The Ethics of
# Molecular Silviculture

PAUL B. THOMPSON

Recombinant techniques for plant transformation in silviculture and agriculture raise ethical issues largely because they have brought the technical practices of breeding before the public eye in a manner that is unprecedented in recent memory. This has placed practitioners in the position of needing to make an articulate and nontechnical statement of the rationale—the ethic—that guides the use and development of science and technology in the plant and animal sciences. Too often practitioners have been unable to muster such a statement. When that happens, the unfortunate result can be an erosion of confidence and trust in the technical competence of specialists and a ratcheting effect that links ethical issues with the perception of elevated risk.

Although it is difficult to propose measures that would constitute a rapid response to this situation in forestry, the longer term need is to develop the capacity to articulate and communicate the professional ethic that guides the technical practices of silviculture and to ensure that technical professionals are receptive to constructive criticisms of their prevailing practices. To that end, in this chapter I provide an overview of normative issues associated with molecular silviculture. Following a brief clarification of terminology, I organize the ethical issues into four categories—religious or metaphysical concerns, environmental ethics, social and political ethics, and professional ethics—and briefly review the contested issues in each category. I include a succinct statement of my own views on the necessity for scientists to participate in ethical discussions as prerequisite for maintaining public trust.

# Terminology

The practices of molecular silviculture relevant to my discussion of ethics include especially the use of recombinant methods for plant transformation in research, production, and conservation contexts but also the development and application of genetic techniques such as genomics and informatics that need not involve plant transformation. Although this definition leaves some ambiguity regarding the scope of the practices under review, ambiguity over the term *ethics* is far more likely to create confusion or misunderstanding. For that reason I will take some pains to clarify the intended scope of ethics and the role that the academic discipline of philosophy can play in elucidating the ethics of any technical practice.

Practices involve ethics insofar as they are understood to serve larger purposes or to be valuable in themselves. To examine the ethics of a practice is to investigate how the practice can be done well or poorly, to inquire into its purposes or value, and to articulate standards for performance, justification, or evaluation of the practice. Some people reserve the word *ethics* for issues involving conflict of interest or sexual misadventure, and the recent spate of interest in research ethics has focused attention on issues of scientific misconduct. The term is also associated with cultural mores, religion, and even irrationality in some quarters.

Philosophers have developed a somewhat more specific interpretation of ethics that stresses the explicit formulation of justifications, desired outcomes, and codes of conduct. Each of these involves a set of claims intended to specify a particular policy or course of action that ought to be followed. Philosophical ethics (or moral theory) is the study of how ethical principles can be systematically used to develop logically consistent and conceptually coherent arguments to support the prescription of a particular course of action. Ethical principles, in turn, are the rules, norms, or specifications that provide the basis for prescribing what ought to be done. The claim that forest policy should promote efficient use of resources, for example, is an ethical principle because it advocates the norm of efficiency as a criterion for the formation and justification of management plans and forest policies.

Philosophers interpret disputes over the legitimacy or justifiability of a given practice as involving competing or incompatible ethical principles. This is perhaps contrary to the assumptions of those who believe that the word *ethics* signals a particular class of special considerations, distinct from those that would be characterized as social or economic. For example, consider a hypothetical dispute between someone holding the view that tree biotechnology is justified because it promotes efficient use of natural resources and someone holding the view that tree biotechnology is unacceptable because it is unnatural. As philosophers see it, this is not a dispute in which only one person is making an ethical argument. Both of these points of view involve ethical principles, and one role of philosophy is to spell out the manner in which conflicting ethical principles contribute to each of these contrasting points of view.

There has arguably been a one-sidedness to press coverage about the ethics of biotechnology. The word *ethics* is generally associated with viewpoints that are critical of the use of biotechnology, often ones that are opposed without exception to all uses of recombinant techniques for plant and animal transformation. Although such critical viewpoints are often countered by sources who cite potential benefits from recombinant techniques, the voices advocating weighing risk and benefit are not represented as expressing an ethical perspective. From the standpoint of philosophical ethics, however, arguing for a practice by citing the relative values of its costs and benefits is a time-honored and logically coherent approach to ethics.

Although philosophers can be expected to use a common vocabulary in discussing ethical issues, it is still the case that the judgments and opinions that individual philosophers develop are somewhat personal. Philosophical literature in ethics often consists in the statement of a particular viewpoint, evaluation, or prescription, and then an argument intended to support the conclusions expressed. What follows in this chapter does not present any detailed arguments. In this context it is important to cover a wide range of issues in a succinct fashion. However, it is also important for a philosophical author to be as unambiguous as possible in communicating the judgments that he or she has reached on the issues in question, even if the presentation does not include all the arguments.

## Religious and Metaphysical Ethics

The first question that leaps forward when the subject of ethics is broached in connection with molecular genetics is whether this whole area of science and technology doesn't transgress some sort of boundary or absolute prohibition. And even if simply learning about the genes is permitted, some clearly believe that moving genes through recombinant techniques is not. This is, in other words, the "playing God" domain of ethics, and it is often associated with the idea that ethics is about misconduct. There is no doubt that many people react to the prospect of altering the genetic makeup of living things with repugnance, and it is not difficult to understand why they might tend to express their reaction by questioning the ethics of such practices. In the case of issues made familiar by press coverage of human cloning and stem-cell research, grounding such reactions in terms of specific religious norms is a fairly straightforward process. However, it is surprisingly difficult to articulate why the alteration of plants, including trees, would transgress generally recognized ethical boundaries or how it would relate to well-established religious traditions.

I can certainly imagine a theological/metaphysical conception of nature that ascribes certain intentions or purposes to a divine being who creates the fabric of reality, and I can imagine that in some theologies or metaphysical systems those intentions and purposes might be inimical to biotechnology. I can also imagine theological and metaphysical views that celebrate humanity's ability to transform nature at the molecular level. I am frankly not the philosopher to offer a sympathetic or deeply felt portrayal of either viewpoint. My reading on the ethics of biotechnology suggests that most people who are inclined to worry about the possibility of playing God are actually somewhat reluctant to make bald statements about God's intentions or purposes. Instead, they refer ambiguously to the sanctity of life or defend the repugnance that many people feel on first learning about the new genetics (see Thompson 1997, 2000).

I do not want to imply anything but respect for people who offer these points of view. In fact, I take great comfort in the fact that they do not profess to be on the hotline to God in deriving their concerns about genetic technology. Nevertheless, my own view is that the humility and caution endorsed by those who take such perspectives are more appropriately expressed as components of environmental or social ethics—issues that will be discussed below. It is not clear why the fact that recombinant techniques are being used warrants a unique kind of theological or metaphysical viewpoint. There is, in my view, a large and growing gap between the language that we use to make sense of the phenomenal world of daily life and the language of molecular biology. The moral wisdom that we derive from our religious and cultural traditions is fitted to a world of rocks, trees, and flowers. Although we should be cautious about discarding that wisdom, it is very difficult to see how it translates to a world of DNA, coding and noncoding sequences, and microcassettes. Those who presume that phenomenally derived norms bearing on topics such as heredity or living appropriately in nature can be applied literally to talk about genes, proteins, and molecular life processes are guilty of naive genetic determinism. Unfortunately, many scientifically trained people are themselves guilty of this—but that is an issue for professional ethics, and I must not get ahead of myself.

Before leaving the religious and metaphysical domain, I want to stress that I am not dismissing these issues. I am not saying that the biologists' language is a true description of reality, whereas ordinary or religious descriptions of the phenomenal world are false. Rather, I am saying that I do not know how to build the bridge between these two kinds of language. I do not know how to apply norms of humility and respect for life at the molecular level. I am therefore not inclined to engage in lengthy discussion of theological and metaphysical ideas. As will become evident shortly, I do think that we can build bridges that relate ethics to some specific environmental and social concerns. I am not sure whether it is important to build bridges in the domain of religious ethics, but if it is, that work needs an articulate spokesperson other than myself.

## Environmental Ethics

There are, I think, many open questions about the environmental risks of transgenic plants, and given the lengthy reproductive cycle of trees, these questions are particularly vexing in the area of silviculture. The main focus of environmental risk from transgenic plants has been the potential for unintended impacts on so-called nontarget organisms: gene flow to close relatives and inadvertent effects on habitat that affect other forms of plant and animal life. Although these are inherently empirical questions, the framing and analysis of environmental risks involve a number of value judgments that require a sophisticated mix of science and ethics. The question of whether to minimize Type I or Type II statistical errors is one example. The question of which populations to specify when formulating relative probabilities is another. Are we interested in transgenic trees as a class? Should they be compared with all nontransgenic trees? Or should we be making a comparison between trees that are genetically similar, save for the transgene of interest? Do impacts on land use count in the environmental risk assessment, even though human decisionmaking would be involved in bringing them about? These are not purely scientific questions, and there should be a more explicit and conscientious effort to address these ethical questions in technical debates about environmental risk.

And then, of course, we get to the question of whether these risks are acceptable. At present, the debate has stressed uncertainty. Does the open-ended nature of these questions provide a reason to block either research or commercialization of transgenic trees? There are environmental activists who think that uncertainty provides the basis for a sweeping argument against transgenic silviculture. They often link their argument to the precautionary principle. Although the precautionary principle can be formulated in various ways, its ethical importance consists in the way that it offers an alternative to norms or decision rules that promote risk-taking whenever expected benefits exceed probable losses. A precautionary approach would differ in that losses associated with environmental damage are treated as a special case on any of several grounds. For example, one might argue that ecological complexity or the relatively weak predictive power of ecological models provides a reason to expect that environmental consequences may be much worse than predicted. The irreversibility of environmental outcomes is also cited as a reason to weigh possible losses much more heavily than expected benefits. In both cases, advocates of a precautionary principle would demand a higher standard of evidence for expected benefits than for possible environmental hazards (see Raffensperger and Tickner 1999).

Although I endorse a precautionary approach in environmental policy, I do not think that it entails a sweeping indictment against biotechnology. The key to my judgment is that even a precautionary approach requires one to evaluate a proposed course of action in comparison with its alternatives. If the alternative to tree biotechnology is that the human species will desist from all use of forest products, precaution might weigh in against biotech. The problem, of course, is that abolishing all human use of forest products would involve such extensive costs that it is not actually a feasible alternative at all. So the alternative to tree biotechnology may actually be an unacceptably exploitative expansion of current practices in industrial forestry. If this is the case, then precaution may actually weigh in favor of transgenic trees. My viewpoint on the environmental ethics of molecular silviculture is that it depends on some background and contextual elements of forest policy, and I need to hear more about it before I could form a firm opinion.

There is also an even more general set of issues in environmental ethics. There has been a tremendous amount of ink spilled in the debate between anthropocentrism and ecocentrism in environmental ethics, a debate often traced back to the philosophical conflict between Gifford Pinchot and John Muir over the future of American forests. Pinchot is portrayed as a figure who saw wilderness as deriving all of its value from the various uses—including recreational uses—that humans make of it. Muir is portrayed as someone who believed that forests, trees, and wilderness were intrinsically valuable, irrespective of any use that humans made of them. Clearly, this debate continues to resonate throughout forest policy (see Norton 1991; Callicott and Nelson 1998).

Does this debate have any bearing on tree biotechnology? My own view is that its bearing is rather indirect, and that certainly the anthropocentrist/ecocentrist divide does not translate directly into pro and con positions on tree biotechnology. Only someone who, taking Muir much farther than Muir himself would have gone, argues against all human use of trees would conclude that absolutely every conceivable application of tree biotechnology is impermissible. Only someone taking Pinchot much farther than Pinchot himself would have gone could think that the impact of tree biotechnology on protected wilderness areas is ethically irrelevant. This brings us back to the issues we started with: the environmental risks of genetic engineering to nontarget organisms. It is certainly possible that different environmental values will lead people to frame questions of risk assessment in different ways, and that is one reason why nonscientists should be included in the process of environmental risk assessment.

## Social and Political Ethics

Many of the activists who have opposed biotechnology in agriculture ground their opposition in a sociopolitical argument. I will sketch the terms of this argument briefly, though I will say at the outset that, in my view, its considerable merits as a case for the reform of our technology policy do not translate into persuasive reasons for singling out genetic technologies. Critics of biotechnology in agriculture allege that it is a tool for increasing the control that a few corporations hold over the entire food system. They see biotechnology as a weapon being wielded against poor farmers in the developing world and as a token in a process of globalization that is intensifying the economic power of multinational companies and international capital.

This is a complex argument in its details, but there are at least four important components to it. One is that the late 1980s through the early 1990s clearly did see a considerable consolidation within major seed, agrochemical, and forest products companies, as well as the pharmaceutical industry. This consolidation was sparked by industry's judgment that genetic technologies would be key sources of profitability in the future and that capturing those profits would depend on vertical integration of technology discovery and delivery processes. Although economists debate whether this consolidation has created monopoly power within the life science industry, the sheer volume of activity in mergers and acquisitions throughout the period could not fail to capture the attention of anyone interested in economic inequality (Teitelman 1989).

Second, U.S. laws on intellectual property changed in the 1980s, allowing for an expansion of patent rights over genes, gene processes, and even whole organisms. This displaced the Plant Variety Protection Act, which was weaker in that it provided exemptions for researchers and for growers propagating plants for their own use. Thus, in addition to industry consolidation, the bigger, consolidated life science companies now had new legal tools at their disposal for concentrating economic power and exerting control (Fowler 1995).

Third, there was a simultaneous shift in the relationship between industry and academic research. In part, this was simply a result of the first two factors. With consolidation among their industrial partners, academic researchers would have found themselves working with a smaller number of firms, even if their actual collaborations with industry had remained unchanged. Academic researchers also found themselves needing industry partners to have freedom to operate in the new era of industrial patents. Some have also argued that the nature of biotechnology has tended to blur the distinction between research and development. The need to acquire patents entered life science departments at universities in a new way, as well, making academic departments seem like private companies to outsiders. The net result was at least the perception that publicly funded, putatively not-for-profit academic science was pretty much indistinguishable from profit-seeking industrial product development (see Kenney 1986; Kloppenburg 1988).

Finally, these events were occurring at a time when economists and sociologists had recently completed new analyses of the way that increases in the efficiency of production technology were linked to socioeconomic changes in rural areas. The so-called technology treadmill was a staple of social science analysis throughout the 1970s. This analysis showed how more efficient production technologies fueled a process of change in the structure of farming, leading to fewer and larger farms. The transition was coupled with a decline in the need for rural service industries and a gradual but inexorable economic decline in rural areas. The theoretical techniques for predicting structural change were applied to some of the early products of biotechnology, and this considerably undercut support for them, particularly among advocates of poor and small farmers. The analysis was also applied retroactively to "Green Revolution" technologies of the 1960s, resulting in a considerable cooling of enthusiasm for productivity-enhancing technologies in the developing world. Again, biotechnology just happened to come along at a time when social scientists were applying these methods to ex ante case studies (see Kalter 1985). Although forestry is different in some respects, it is not wholly different, and the timing of new biotechnologies coincided with a new level of consciousness among economically disadvantaged producers about the effects of technology on their interests.

The combined upshot of these four factors was that biotechnology became a cynosure for those who see technology as a force driving modern societies toward economic and political inequality. Despite thinking of myself as a political moderate, my own view on the social ethics of technical change is actually very close to that of Andrew Feenberg, who describes himself as a left-leaning pro-socialist critic of capitalism (1999). Feenberg believes that technological innovations have indeed tended to serve the interests of capital throughout history, but he also believes that this has primarily been because of the way that owners of capital have been linked to the developers of technology through social networks. In most cases, it would be possible to have the benefits of new technology without the socially destabilizing inequalities, if only the developers of technology could be linked in networks with comparatively disadvantaged people. Thus, what is needed is a political reform of the social infrastructure for developing technology, not opposition to any particular form of technology itself. Although I have never thought of myself as pro-socialist, and though I think the details of any reform will prove to be pretty complex, I find myself in substantial agreement with Feenburg's social ethic for technology.

Yet none of this really provides an argument against tree biotechnology. The implication is that those who are opposing biotechnology for reasons of social ethics are chasing the cape, when they should be after the bullfighter. I must qualify my remarks by saying that I do not want to ban corporations or the profit motive; nor do I want government ownership and control of technology development. As I have said already, the details will be complex. But in each of the four elements described above, it is the social networking, far more than the use of gene-based or recombinant techniques, that leads to the unfortunate social results. We could put an end to biotechnology tomorrow, and it would not improve the situation with respect to economic inequality one iota.

## Professional Ethics

And this brings me to my final domain. For almost 100 years, the professional ethic in the life sciences has been to avoid ethical issues when at all possible. Life scientists came by this ethic honestly and for good reason. The ideal of scientific objectivity became crucial to the establishment of rigor and credibility in scientific disciplines. Entanglement in religious debates over evolution and abortion came to seem less and less relevant to the conduct of science on a day-to-day basis. Nevertheless, by absenting themselves from any discussion of the social networks in which their work is applied and the technologies that are adapted from it, life scientists have adopted an ethic that permits powerful actors to use science silently in the extension and exertion of their power. The uses to which science has thus been put are often quite defensible and have in many cases been progressive. Yet even in those cases, I submit that the quietness of the alliance between science and economic power is distressing.

Ironically, the quietness of the life science community is coming to undermine the very objectivity and credibility that a previous generation's professional ethic was designed to ensure. There is, I submit, a feedback mechanism between the quietude of life scientists with respect to the social implications of technology and the public's willingness to place confidence in their opinions with respect to environmental risk. Environmental activists are networked with social activists in attempting to constrain the growth of global corporate power. They are bound to overhear some of the indictments raised against biotechnology in those quarters. The silence of the life science community with respect to such issues can be deafening. The next inference is unfortunate, in my view, but not altogether unexpected. It is implicit in the question, How can people who are so closely allied with corporate interests when it comes to social issues be trusted when it comes to the environment? And then there is one more inference that is made: Isn't it dangerous to leave the future in the hands of such people? In this way, the silence of life scientists on social issues is translated into positive allegations of environmental risk (Thompson and Strauss 2000).

Now I must be as clear as possible. I do not believe that life scientists as a group are in league with corporate interests; nor do I believe that corporations are evil, or even that their interests are inherently antithetical to the social ends that I support. I certainly do not believe that life scientists' failure to become involved in debates over the social control and socioeconomic impact of biotechnology contributes to environmental risk. What I am saying is that these are fairly natural inferences for people to make. And I believe that such inferences have substantially frustrated the accomplishment of both environmental and social aims that I strongly support. The expenditure of goodwill and intellectual resources in opposing biotechnology is, in my view, a perfect waste of energy and money by people whose general values and aims I endorse. I wish that I could effectively make the case for my view among environmentalists and social activists. But I am also saying that an ethical failure among life scientists has contributed to this unfortunate result, as well.

Science and technology do not automatically translate into socially beneficial outcomes. If the outcomes are to be beneficial, conscious and deliberate work must be done to build the social networks and to think through the environmental impacts. Undertaking such conscious and deliberate work is, in my view, a needed and too often lacking element in the professional ethic for the life sciences. The 2001 symposium that originally generated the chapters in this book is a notable and important exception to the general trend, and I hope that my remarks in this chapter will be taken as encouragement to follow through on the work that was begun there.

## Conclusions

When ethics is understood as making explicit the rationale for why something should or should not be done, and being willing to defend that rationale against challenge, then we can say that biotechnologists in all fields could do a better job with respect to ethics. Often the reasons for using transgenic technology are left implicit, and challenges to its use are met with silence. It is even worse when reasonable and understandable (but often answerable) challenges are characterized as irrational or scaremongering by reputable scientists. There is a need for ethical debate over the application of molecular transformation and even genomics technology to silviculture and a corresponding need to build a record of careful and explicit arguments on the rationales, concerns, risks, and justifications for tree biotechnology. The responsibility to engage in such debate should come to be seen as part of the professional ethic for molecular silviculture.

The substance of these debates can be expected to involve religious or metaphysical issues, environmental impact, and the social context of applied forest science and forest policy, as well as the emerging professional codes of practitioners. Although it is impossible to review all the particulars in a short essay, my judgment is that there are not, at present, persuasive philosophical or risk-based arguments that justify prejudicial policies, much less a halt to research, on tree biotechnology. There are uncertainties that deserve both study and debate, and the eventual approval and commercialization of forest biotechnologies will require clearing a number of specific hurdles. Those who hope to use molecular techniques must realize that a continuing failure to address the empirical questions pertaining to risk or to build a record of responsible ethical debate will eventually warrant the conclusion that the science should be stopped because its practitioners cannot be trusted. If uncertainties and ethical silences are allowed to linger, the public will be entirely justified in regarding tree biotechnology with the same skepticism and hostility that now plague agricultural and food applications.

# References

Callicott, J.B., and M.P. Nelson. 1998. *The Great New Wilderness Debate.* Athens: University of Georgia Press.

Feenberg, A. 1999. *Questioning Technology.* London: Routledge.

Fowler, C. 1995. Biotechnology, Patents and the Third World. In *Biopolitics: A Feminist Reader on Biotechnology,* edited by V. Shiva and I. Moser. London: Zed Books.

Kalter, R.J. 1985. The New Biotech Agriculture: Unforeseen Economic Consequences. *Issues in Science and Technology* 13: 125–133.

Kenney, M. 1986. *Biotechnology: The University–Industrial Complex.* New Haven: Yale University Press.

Kloppenburg, J. 1988. *First the Seed: The Political Economy of Plant Technology: 1492–2000.* Cambridge: Cambridge University Press.

Norton, B.G. 1991. *Toward Unity among Environmentalists.* New York: Oxford University Press.

Raffensperger, C., and J.A. Tickner. 1999. *Protecting Public Health and the Environment: Implementing the Precautionary Principle.* Washington, DC: Island Press.

Teitelman, R. 1989. *Gene Dreams: Wall Street, Academia and the Rise of Biotechnology.* New York: Basic Books.

Thompson, P.B. 1997. *Food Biotechnology in Ethical Perspective.* London: Chapman and Hall.

———. 2000. *Food and Agricultural Biotechnology: Incorporating Ethical Considerations.* Ottawa: Canadian Biotechnology Advisory Committee.

Thompson, P.B., and S.H. Strauss. 2000. Research Ethics for Molecular Silviculture. In *Molecular Biology of Woody Plants,* vol. 2, edited by S.M. Jain and S.C. Minocha. Dordrecht, Netherlands: Kluwer, 485–511.

# 8

# Will the Marketplace See the Sustainable Forest for the Transgenic Trees?

DON S. DOERING

This chapter explores the critical lessons that the commercial introduction of genetically engineered food and fiber crops provides to the nascent tree biotechnology industry. Extracting these lessons from the still unfolding commercialization of genetically engineered crops is a subjective process, and projecting them to forest biotechnology is necessarily speculative. However, the agricultural biotechnology experience suggests that the future of a publicly or privately financed market for genetically engineered (GE) trees depends on how the technology is first applied, who benefits from the introduced traits, who participates in the decisions about commercialization, and society's acceptance of tree plantations and the release of genetically engineered organisms into the environment. The introduction of genetically engineered trees to the market will best occur via an appropriate regulatory framework and a collaborative effort of the public and private sectors, including stakeholders from civil society. Society will only accept GE trees if the products have broad social utility, are designed for environmental safety, and are well tested for ecological impacts.

Today we are far from the goal of forests and tree plantations' sustainably providing all the goods that the world wants—from timber to ecosystem services such as biodiversity, protected habitat, watershed protection, and carbon storage (WRI 2000). Almost all of recent deforestation has occurred in the highly biodiverse forests of the developing world, whereas most reforestation and afforestation have occurred in the developed world (Matthews et al. 2000). Though fiber production has risen in the last decade, net global forest cover has dropped by 9.4%, with deforestation (−14.6%) slightly offset by reforestation (+5.2%). Some of the increase in forest cover is from plantation establishment, which now accounts for about 5% of global forest cover, and plantations supply about 35% of industrial roundwood (FAO 2000). About half of the annual consumption of trees is for the fuelwoods that are the primary energy source for about two billion of the Earth's people (Matthews et al. 2000). Society needs an increasing and sustainable timber supply, and it needs healthy, biodiverse forests.

People have always applied technologies to meet their needs for forest products, including both sustainable and unsustainable forestry and land use practices. Genetic engineering is no more inherently sustainable or beneficial than chemical, civil, or mechanical engineering. What matters is how and to what traits and species we apply it. The eventual societal impact of genetically engineered trees depends on how the technology is used, how economic markets develop for wild and domesticated trees, and how those markets are linked by supply, demand, and public policy.

## GE Trees: A Question of How and When

The discussion in this chapter reflects my own judgment that genetically engineered trees will someday be commercialized but that the timing, impacts, and trajectory of commercial introduction are far from certain. There *are* conditions under which transgenic trees can reach the marketplace and can serve economic development as well as social and environmental interests. For that to happen, however, calls for an approach to the commercialization of genetically engineered trees that is significantly different than what has occurred with the first generation of genetically engineered crops. Will a market develop in some countries, while intense societal opposition closes markets in others, as has been the case for GE food crops? Can we learn from experience with other genetically modified organisms to improve the chances that public and private genetic engineering of trees will safely and fairly serve the needs of society? Will genetic engineering be applied to trees to address critical global needs, with respect for societal concerns, or only to serve the narrower interests of developed-country corporations? The financial allure of the new technology and of high-technology solutions to forestry problems will make it hard to keep sight of sustainable forestry goals beyond the transgenic trees.

Genetically engineered trees are in the environment. In hundreds of experimental field trials of dozens of species, and in several countries, private and public sector scientists are conducting research on transgenic trees. The U.S. government has received applications to test 138 transgenic trees, 58 applications in the last two years (Peña and Sequin 2001; Mann and Plummer 2002). None of these are precommercialization trials in a regulatory sense, although many may be viewed as commercial prototypes. There is cautious interest within the forest industry, and distinct interest among academic scientists, in genetically engineered trees, as publication of this book attests. Some of the same companies involved in food crop biotechnology are involved in GE trees, and many field trials of trees are testing the same input traits of glyphosate tolerance and *Bacillus thuringiensis* (*Bt*) endotoxin expression that are in the commercial GE crops. The proposed benefits of tree biotechnology, such as greater yield, meeting global needs, lower production costs, and natural resource conservation, share many similarities with the aims of industrial crop biotechnologists. These similarities are reminiscent of the early phase of the agricultural biotechnology industry, and it appears inevitable that GE trees will be commercialized someday.

Genetically engineered trees are intriguing to business, but it is hard to imagine that a forest industry executive or a tree biotechnologist would want to see the product introduction of trees follow the precise example of GE food crops. The introduction of herbicide-resistant and pesticidal crops is a story of fantastically fast technology adoption and market penetration. Only five years after their introduction, almost three-quarters of U.S. cotton, over half of U.S. soybeans, and a fifth of the U.S. corn crop are planted in GE varieties (Carpenter and Gianessi 2001). Despite this apparent success, however, GE food crops have been a product introduction nightmare. Agricultural biotechnology executives have been fired or replaced, corporate reputations have been shaken, billions of dollars of shareholder value has been lost, and agricultural biotechnology divisions of large companies have been swapped and expelled from their parent companies (Challener 2001). The time and cost required to achieve regulatory approval and the cost of import/export permissions continue to rise rather than decline, as new regulations are enacted for commodity segregation, labeling, environmental testing, and food safety testing. The unity of Europe in rejecting biotechnology crops has slammed closed valuable grain markets, and opposition to "Frankenfoods" has become a rallying point of the antiscience, anticorporate, and antiglobalization movements (Moore 2001). As we contemplate increased corporate and government investment in tree biotechnology, it is fair to wonder how to ensure that society will benefit. Will there be corporate casualties and social polarization along the way? And what is the best path forward?

## Listening to the Lorax: Creating a Twenty-First-Century Market

*The Lorax* is a clear and compelling story of unsustainable forest management and unsustainable business strategy (Seuss 1971). In this classic children's story by Dr. Seuss, the Once-ler uses the fiber of the fictional Truffula tree to knit thneeds, a multipurpose textile that is "a-fine-something-that-all-people-need." The demand is immediate and voracious, and it is served by the Once-ler's innovation, which automates thneed production and Truffula tree harvesting over the repeated and furious objections of the Lorax. The Lorax is himself a forest resident and the spokesperson for other Truffula ecosystem stakeholders such as the Bar-ba-loots, Swomee Swans, and Humming Fish. As the thneed industry and automated tree harvesting degrade the basis of the ecosystem, the native species migrate, the resource is depleted, and the thneed business collapses. The Once-ler is left in financial ruin and becomes—too late—an advocate of ecosystem restoration.

Were Dr. Seuss to invent this story today, he might incorporate the powerful trends that are shaping the twenty-first-century marketplace (Doering et al. 2002): powerful, multinational corporations, markets strongly influenced by corporate intentions and political influence, technologies being pushed to the marketplace. However, a product's performance now includes not only consumer need, but also the environmental footprint of its manufacture, regulations for product life cycle stewardship, recycling, and environmentally friendly design. The ability to sell to society is not only governed by demand; society must also grant license-to-operate (Elkington 1997; Arnold and Day 1998). Many corporations have been forced to change business practices and strategies because of civil society activism, among them some of the largest timber companies and wood buyers (Gereffi et al. 2001). Social license-to-operate, societal acceptance, stakeholder engagement, and corporate social responsibility are now central strategic issues of business leadership and not just public relations functions. Social and environmental issues are not only matters of marketing but are primary considerations for scientists and engineers at the first moments of product design (Tischner and Charter 2001).

If *The Lorax* were to be updated today, the Lorax would be supported by outside activists in his opposition to the Once-ler; Bar-ba-loots would be found chained to Truffula trees and to thneed shop shelves; legislation would protect the Swomee Swans. Truffula forests might be certified for sustainable harvesting, and perhaps the story would have a happier ending. At the beginning of the twenty-first century, the voice of the Lorax is louder than ever. The European rejection of America's genetically engineered crops richly illustrates that people and interests other than customers and shareholders now have the power to close global markets through protest, political power, and boycott (Moore 2001). Sustainable business principles should guide forest biotechnology advocates to engage with a broad range of groups and individuals who define themselves as stakeholders in forest and tree-related issues to ensure that product designs, applications, and benefits serve an array of human needs.

# Crop Biotechnology's Failure to Earn Public Trust

## Unifying the Opposition

The commercial introduction of genetically engineered crops unified many social movements to an unprecedented degree and armed them with potent symbols with broad public appeal, such as baby food, monarch butterflies, and the small-scale farmer (Lenzner and Kellner 2000). A common cause was found that brought together individuals and nongovernmental organizations with interests in food safety and consumer choice, farmers' rights and property rights, spirituality in a technological world, economic justice and global trade, curtailing corporate influence, combating hunger, and protecting the environment (see, for example, the campaigns of the Genetically Engineered Food Alert, www.gefoodalert.org; the Organic Consumer Association, www.purefood.org; and the Five Year Freeze, www.fiveyearfreeze.org). There are underlying themes to the opposition to crop biotechnology that merit deep consideration by a nascent forest biotechnology industry: individual choice, product safety, social utility, transparency, and ethics.

The first products of the biotechnology industry were seen (and still are seen by many) as being potentially unsafe for humans, animals, and the environment, and the utility of the products to the consumer and even to the farmer is questioned. The transparency of companies (or lack thereof) has been under fire, as have been the pro-biotechnology stances of the U.S. and European governments and the lack of transparency in the regulatory process for GE crop approvals. Opponents of biotechnology regard with great distaste the lack of personal choice about foods derived from genetically engineered commodities and what they perceive as secretive and unethical actions by the biotech companies in imposing the products on the market. Both corporations and regulators have responded to these concerns and charges, but the strongest opponents of genetic engineering have not tempered their opposition (Victor and Runge 2002). The positive changes may be fully institutionalized by the time the first mass-market GE tree approaches the market, and there may be better established ways for tree biotechnology companies to address public concerns.

## Distrust of Corporations

The benefits of transparency, respected ethics, and the support of social values are encapsulated in the word *trust*. The agricultural biotechnology companies, and by association many scientists who worked with them, have largely lost the trust of the politically active public, particularly the "anti-biotech" activist organizations. The loss of trust has many origins, with lessons for commercialization of GE trees. The actions of automotive, tobacco, pharmaceutical, chemical, and oil companies have made people suspicious when large and apparently powerful companies claim loudly to be acting for the public good, providing safe and beneficial products, while they simultaneously fight regulations related to human and environmental safety and liability for past damages. The logging and timber products industry has similar reputational liabilities for its poor past and current environmental record and political stands and activities. Images of clear-cuts on public lands and protesters blocking logging trucks have been vivid symbols over the past 30 years of the environmental movement. The logging industry was a strong backer of the Bush administration's December 2002 action to greatly reduce the environmental impact assessments required for logging in U.S. forests—a change that conservationists widely decried (Coile 2002). A union of the timber industry and the plant biotechnology industry comes poorly armed to a battle based on public trust and may be a dream marriage to an anti-corporate activist (Lenzner and Kellner 2000; Sampson and Lohmann 2000).

## A Critical General Public

The problems that the agricultural biotechnology companies encountered came not just from opposition activists, but also from a public sympathetic to the activists' concerns. Defining what members of "the public" think and who they are is not easy; nor can one perfectly predict what the general public will think in the future. In a 2001 National Science Foundation poll of the American public, 61% of those surveyed placed themselves in response groups that thought genetic engineering's benefits equal its harmful results (28%); that harmful results slightly outweigh benefits (19%); or that harmful results strongly outweigh benefits (14%) (NSB 2002). The total of these three groups has been growing since 1985. Similar numbers were found in a January 2002 poll of American adults by the Pew Initiative on Food and Biotechnology, in which 38% of those surveyed believed that the risks outweighed the benefits of genetic modification of "plants, animals, fish or trees" and 21% considered the risks and benefits to be about the same (Pew 2002). Europeans and Canadians are considerably more negative than Americans on biotechnology (Gaskell et al. 2000). These and other surveys also suggest, however, that there is a significant portion of the public who are likely to be suspicious of genetically modified trees but will consider the specific benefits and risks of each case (Priest 2000).

Genetic modification itself is controversial, and yet we have examples showing that society weighs its risks against its benefits in particular cases. That is likely to be the case for trees, as well. The genetic engineering of microbes to produce vital medical therapeutics, such as insulin or erythropoietin, is generally considered acceptable because the GE organisms are under strict containment and unable to replicate in the environment, there is general confidence in the regulatory system, the product has a clear utility in saving lives, and the medical consumer has information and choice (NSB 2002). Those conditions constitute a high standard to reach, but they hint that environmental safety and containment, high social value, and choice might someday describe acceptable GE trees.

An interesting emergent case is the production of pharmaceuticals in plants, or "biopharming." In making the technically expedient choice of fertile corn—a commodity food crop—as its genetic background, the pharmaceutical industry is taking a very high-risk approach to ensuring environmental containment and segregation from food and feed. Even in its field trial stage, the predicted issue of contamination has already arisen, when soybeans were planted over a field that had been used by the biotech company ProdiGene (College Station, Texas) for a field trial of corn expressing an industrial chemical compound. The soybeans became contaminated with the corn, leading to the destruction of 500,000 bushels of contaminated soybeans and costing the company a $250,000 fine plus the cost of the destroyed soybeans (Pollack 2002). The battle lines are being drawn over regulations, and this case may help reveal the extent of public tolerance for balancing the social utility of pharmaceutical production with the risks to food safety and the environment (Freese 2002; BIO 2002).

### Aligning Messages with Actions

The agricultural biotechnology industry used, and still uses, the promised benefits of biotechnology for sustainable food supply to earn its social license-to-operate (e.g., CBI 2002). In doing so, the industry has picked an unfriendly and difficult arena in which to take on opposition interests, for it has little to show that the current products meet the promise of the technology. First-generation biotechnology products did not address causes of food insecurity and indeed were never designed to be grown in the climates and agricultural systems of regions where food insecurity exists. The stated aims of food security and sustainable agriculture put the industry in debates with opponents who have superior knowledge of global food needs and who can produce harsh critiques of industrialized agriculture and its impact on the environment, international food trade, and the farmer. The agricultural biotechnology companies' relative financial investments in "public good" projects versus industrial agriculture are not proportional to their treatment in the industry's public relations material. Such dissonance between message, image, and action fuels distrust.

To earn public acceptance, it is important to be quite clear about the hypothetical benefit of the technology and the risks and benefits that the public and the market are being asked to accept. Forest biotechnology will not save the world's forests in coming decades, any more than crop biotechnology will solve problems of food insecurity; specific *products* may make positive contributions. Benefit to society is not an inherent property of the technology but is a result of choices and investments in how technology is applied, who benefits, and who is involved in decisionmaking. Technology is only a small component in the solution to agricultural problems that are often rooted in long-standing historical, economic, and political inequities, corruption, particular resource ownership regimes, environmental misman-agement, agricultural production subsidies, policies that do not internalize environmental costs, and traditional harvest practices.

The promised future benefit of GE trees to society cannot be used to jus-tify the commercialization of products that do not deliver those benefits. There has also been little or no detailed analysis to propose how genetically engineered trees might directly relieve pressures on biodiverse forests that are threatened or might significantly affect fuelwood, timber, and pulp supplies in the regions of highest future demand. The current analyses offer an approach to the problem, but they have not yet been applied with regional and national specificity (Fenning and Gershenzon 2002; see also Sedjo, Chapter 3). A detailed analysis must consider fully the costs of product devel-opment, regulatory approval, regulatory enforcement, and monitoring, as well as how the genetically modified trees compare with the natural resource and non-GE trees in availability, quality, production cost, and ownership. Such analysis can provide the basis not only for communicating trustworthy messages to the public, but also for formulating public policy and for scien-tific research.

## *Maximizing Social Utility and Sustainability*

Sustainable business is business that raises its social utility by creating envi-ronmental, social, and economic value, while not depleting resources. Good environmental performance in the past meant doing "less harm." Today's stakeholders demand performance beyond regulatory compliance and favor companies and products that do "more good" rather than merely less harm. Many have observed that there might have been greater societal acceptance of biotechnology if only its first products had yielded direct consumer benefits, such as better tasting, more attractive, more convenient, safer, more nutri-tious, or cheaper foods (Victor and Runge 2002). But the agricultural bio-technology industry has asked the public to bear unknown risks from genetic engineering with no perceived direct benefits. The same desire for direct ben-efit and minimal risk will also apply to genetically engineered trees and their products.

The virus-resistant papaya case illustrates public acceptance of a geneti-cally engineered fruit tree whose traits created direct local value. The GE papaya (*Carica papaya*) is the only genetically modified tree that has been approved and is in commercial production in the United States. Introduced in 1998, it is resistant to the papaya ringspot virus (PRSV) and was developed collaboratively by two trusted universities with highly regarded agricultural programs, Cornell University and the University of Hawaii. Opponents of genetically engineering trees include the papaya in their opposition, but it has not gained importance as a symbol or a leverage point for anti-biotechnology groups.

There appear to be several features of this successful commercial introduction of a GE tree that might be models for the future. First, the tree resulted from a public collaboration, making the processes, science, data, scientists, and goals relatively transparent from the project's outset. Second, it produced an economic benefit that was directly appreciated by the local community: the saving of Hawaii's papaya industry. Third, there are no close, wild relatives of papaya on Hawaii, and the island is isolated from the center of origin. Fourth, the GE papaya did not involve the use of chemicals or a chemical company. Fifth, the modification was with a coat protein of the virus and can be understood as a "vaccination," with no suspected ill effects on insects or wildlife. Lastly, it worked. Hawaiian papaya production, which had been in long-term decline because of the PRSV epidemic, grew 35% from 1998 to 2000, as just over 50% of the crop was converted to the GE variety. (NCFAP 2002).

The lesson for business is that it ought to transparently test, quantify, and communicate the economic benefits of tree biotechnology that accrue both to the company's interests and to society's. The typical consumer understands the desire of businesses to grow, to reduce costs, and to eliminate environmental liabilities. Genetically engineered tree plantations in regulated and industrialized markets may very well improve corporate profitability and reduce a company's net environmental impacts. "We're doing this for our business" will be accepted more readily by a skeptical public than will claims that massive investments in new industrial technologies are motivated by desire to serve the local public good or to save threatened forests in distant markets. Developing countries have very different problems and priorities than the developed world, and meeting their needs requires specific scientific and technological priorities (Weil 2001; Daar et al. 2002; Huang et al. 2002b). A transgenic pine in Georgia will no more save the forests of Indonesia than an improved soybean grown in Iowa will benefit the food-insecure peoples of Africa and Asia, if those markets are not economically linked. To build public trust, tree biotechnologists will need to match investment with oratory and choose messages that are simple and true.

People invited to live next to genetically engineered forests or orchards will ask, Are we getting wood or paper that is better, stronger, longer-lasting, more appealing, or cheaper? or Are we getting fruit that is better-tasting, longer-lasting, more nutritious, or cheaper? Considering the lack of differentiation among so many timber and paper products and their low cost, the products of genetically engineered trees may not deliver these kinds of direct consumer benefits. With harvest and farm costs (plantation costs) but a small component of retail prices, and with the costs of development, regulatory approval, and compliance high, it seems unlikely that products of genetically engineered trees will be cheaper. What genetically engineered traits in trees, then, can create benefit for those in society who may bear unknown environmental impacts or be asked to absorb the costs of development and regulation?

## Traits for Publicly Supported Forest Biotechnology

One of the challenging issues associated with agricultural biotechnology has been the respective roles of the public and private sectors and the fair expectations of each sector to deliver benefits to society (Smith et al. 1999). Deployment of a strong, publicly financed research effort and nonproprietary technologies applicable toward public goals is in the interests of both society and business. A market for tree biotechnology need not be an unsubsidized market but might be developed through the significant public investment in tree biotechnology. What if publicly funded agricultural biotechnology had preceded the private sector's rush to market? It is easy to imagine better engineered products, a public more receptive, and a market more open to genetic engineering if the first genetically engineered crops we had heard about were public projects such as vitamin A-enhanced rice, a virus-resistant sweet potato to feed Central America's hungry, or a high-protein cassava that grew in the depleted soils of East Africa.

Such a "public-first" scenario is still possible for tree biotechnology, though it will require large investment and careful choice of target species and preferred traits. The first public priority may be rapid reforestation of abandoned and degraded agricultural lands to create measurable benefits of soil stabilization, watershed protection, habitat restoration, and timber production. Fast-growing plantation trees designed for tropical zones might also be used to create plantation buffers around threatened tropical forests to supply pulp, timber, fuel, and forest products to local communities. Fast-growing fuelwoods that grow on marginal soils might also help protect forest frontiers, raise living standards, be a component of carbon sequestration markets, and support economic development.

Another development that would facilitate public acceptance of genetically engineered trees would be engineering specific disease resistance that saves a tree of high environmental, economic, or symbolic value. In America, genetic engineering for fungal resistance that would allow the restoration of the American elm, Fraser fir, or American chestnut to Eastern forests could have a large positive impact on ecosystem restoration and on the tourism, landscaping, timber, and forest product industries (Mann and Plummer 2002).

A wide variety of projects are under way in the public research sector on reduced or increased lignin content, increased cellulose content, faster growth, more uniform growth, growth in marginal or arid soils, and other traits. Because the investment in these projects is small and they concern the plantation species of the developed world, they may not result in the high-impact, "icon" products suggested previously, which would shape societal opinion and drive new research and policy. More fanciful possibilities include engineered control of stress response and adaptation, to allow response to climate change; production of biofuels; and trees specifically designed to maximize carbon sequestration.

In advocating a strong public effort in tree biotechnology, I do not merely offer an industrial strategy for public acceptance. It is my hypothesis that a public effort will be better directed to public goals, will be more transparent, and will create more common knowledge and common intellectual property than private and proprietary efforts. Public sector efforts to address public issues in tropical areas through tree biotechnology face the same challenges as the application of agricultural biotechnology to global food needs: a relative lack of scientific knowledge of tropical species, lack of scientific and regulatory capacity in developing countries, lack of regulatory enforcement capacity in developing countries, high diversity of species and culture methods, and the concentration of research and development dollars and intellectual property in the private sector. The tree biotechnology research agenda today does not appear to be driven by a global analysis of the needs for trees and forests. Creating and commercializing a genetically engineered tree to deliver measurable public benefit would call for hundreds of millions of dollars in public and private investment in science that is guided by a deep needs analysis and by public participation in decisionmaking.

## Traits for Industrial Forest Biotechnology

Among the traits and benefits that forest biotechnology is interested in, the reduced-lignin designs may have the greatest social utility; they would reduce chemical, water, and energy use by the pulp and paper industry, as well as the pollution it creates. Though the cost savings may not reach far down the value chain, the benefit will occur at the site of milling: People who live at or near the plantations and the paper mills would enjoy cleaner water and air. These "green" properties may even allow branding of an otherwise undifferentiated product. But there is a catch: Selling cleaner processes involves admitting that current processes are dirty and are environmental liabilities. How much of a paper company's resource base has to be from reduced-lignin trees to make a visible environmental impact sufficient to outweigh subjecting all of its production to scrutiny? As air and water regulations and energy costs put increasing pressure on the industry, the benefits it could reap from reduced lignin will increase, as will the competitive advantages and financial value of cleaner processes (Repetto and Austin 2000).

A second trait with high industrial and social utility is fast growth. In this case the goal would be to develop a fast-growing pulpwood or hardwood with a short enough rotation time to make plantations less costly than logging in biodiverse frontier or secondary forests. In developing countries, fast-growth plantations might increase fuelwood production and remove forest areas from logging if deployed in a policy context that protected native forests. Another major industrial, and perhaps environmental, benefit of fast growth would be to reduce the generation time of marker-assisted, selective breeding of trees. As in the lignin example, benefits would need to be regional, if not local, to those bearing any risks. A company may be able to demonstrate to the public that its fast-growing GE tree plantation eliminates its need to log on public and private lands or to purchase wood from threatened forests. Such claims would be much more convincing if proved by the company's selling or giving private forestlands to the state for public use and providing a transparent accounting of its timber sources. Appropriate public policy would need to be in place to ensure against a perverse consequence, such as the devaluation of forests as timber repositories, leading to their destruction for other land use or material needs. Alteration of fundamental properties such as lignin content and growth rates will also demand mechanisms to guarantee no flow of potentially deleterious traits to wild tree ecosystems.

It will be hard for a forest biotechnology industry to justify genetically engineered tree plantations in places such as New Zealand, Canada, Europe, or the United States based on claims that they relieve pressure on threatened forests such as those of Russia, Indonesia, the Amazon, or Gabon. We are still not in a situation of scarcity in timber supply—either due to resource depletion in most areas or as a result of public policies—that reflects the threats to forest ecosystems (FAO 2000; WRI 2000). Improved timber supply in highly developed countries may have little impact on distant markets. Or it may act to lower prices and drive greater natural resource depletion by encouraging destructive, low-cost harvesting methods to maintain incomes. A comparative analysis in industries such as mining, fisheries, and grains may identify the conditions under which new production methods do or do not affect resource exploitation. Even in a trade as liquid and industrialized as grain commodities, oversupply in the United States has neither stopped ecosystem conversion to agricultural use nor fed the rural poor in distant developing nations (WRI 2000). Agricultural biotechnology has revealed the different cost–benefit considerations of new technologies in different parts of the world. A significant Asian and African voice has emerged calling for access to biotechnology and self-determination of its application to food security, even as European and developing-country activists have sought to prevent genetic engineering in tropical regions (Juma 1999; Wambugu 1999). Determining what policies create the right market scarcity—that drives conservation, appropriately values forest ecosystem services, and links international timber markets—is an important research priority if we are to model accurately the impacts of GE tree plantations and the benefits of traits such as fast growth.

A plantation of fast-growing trees in Asia that saves adjacent forests and meets local fuelwood, timber, and pulp demands is a much more direct value proposition than the assertion that GE trees in the United States will have a beneficial impact on distant markets. Well-enforced policy that protects forests is a likely prerequisite for market development that favors both plantations and high-technology interventions. After over half of its original forests were lost, it has taken 75 years of strict government policy for New Zealand to build a plantation industry that can sustain its timber product needs. In the case either of reduced-lignin or fast-growth trees, a private tree biotechnology company will have to meet the highest standards of environmental safety, ethics, and transparency to win the public trust. The important qualification of environmental safety is most likely to be met by multiple, benign mechanisms for tree sterility and by plantations that are not developed at the expense of native forests and that are managed for biodiversity and ecosystem services.

## Design for the Environment

The major determinants of a market's developing for GE trees will be the environmental safety profile of the proposed products, the regulatory costs of safety testing, and societal acceptance of that safety assessment. Most agricultural molecular biologists do not label themselves as genetic engineers, and the language of engineering and design is not used to describe genetically modified crops. However, these crops *are* engineered products, and an engineering mindset may serve both the industry and society. Engineers have spent the last several decades learning and proving that environmental benefit is best achieved by design and that approximately 80% of the environmental impact and cost of a product is determined at the point of design (Tischner and Charter 2001). The end-of-pipe solutions of scrubbers, waste treatment, and toxic material disposal are far more costly to society, business, and the environment than pollution prevention at the moment of design. The same is true of genetically engineered crops, and the same will be true of genetically engineered trees.

An informal mapping at World Resources Institute of the environmental impacts pertaining to genetic engineering of food crops sorts them into two major categories: (a) direct impacts, such as compromised human food safety, ecosystem harm, harm to animals, loss of genetic diversity, and resource depletion, and (b) unknown and indirect impacts that GE crops may have on ecosystems and on the intensification and spread of industrialized, chemical-intensive monoculture. At least four primary mechanisms may mediate most of the potential direct environmental threats: toxin production, gene disruption, weed creation, and genesis of new pathogens. At the root of these four mechanisms—and indeed of all the concerns—appear to be three core issues: the control of gene expression, the potential for gene transfer, and the intended design of the engineered organism.

The agricultural biotechnology industry is just coming to appreciate the implications of design and the analogy of front-of-pipe designs to reduce cost and risk. Consider, for example, that the design of *Bt* corn was simply to achieve gene expression in corn. The goal accomplished, that is, constitutive expression of *Bt* toxin in all corn tissues, among them the corn pollen, has led to the high costs of testing pollen flow; the need for extensive refugia and complex grower contracts and compliance schemes; contamination of adjacent, certified organic corn; and the persistent controversy over impacts on nontarget lepidopterans such as the monarch butterfly. Another example of design implications is the need to eliminate antibiotic resistance markers and to develop alternative, selectable markers, as reflected in the new European Union biotechnology regulations as well as public commitments by Monsanto Corporation (EU 2002; Monsanto 2002). These cases suggest principles for design, such as (a) that the product of an introduced or modified gene should only be specifically released into the environment; and (b) that the only functional open reading frames in transformants will be the gene of interest.

The performance, risks, and benefits of first-generation biotechnology products may have been very different had such design-for-environment principles guided the priorities of basic and applied research and been used to stimulate innovation. Tree biotechnologists can adopt the mindset of green product designers and use design principles for environmental safety as they develop new products and set the frontiers of basic research. The transgenic trees planted to date for research purposes should be recognized by their creators as the experiments that they are and should not be confused with product prototypes or with products engineered for the marketplace and the environment.

## The Problem with Input Traits

Though it is easy to speak in general terms, a GE tree's environmental impacts, commercial benefit, and social acceptance are specific to the engineered trait it carries and the physical and cultural context of the silvicultural system. Public opinion surveys in both the European Union and the United States indicate that the public accepts different genetic engineering applications to different degrees (NSB 2002). This point cannot be overemphasized: the engineered trait (i.e., the modification and its expression, whether in the plantation or in an unintended context) is ultimately the source of measured and perceived social utility. The first crop biotechnology products in large-scale planting are corn, cotton, soybean, and canola. All feature "input" traits, meaning traits that act as production inputs, or work in conjunction with production inputs, to the agricultural system, and their benefits accrue to the supplier of the input and to the farmer. The marketed product has no new functional characteristics for the end customer.

Only a small volume of crops such as soy, canola, corn, and even cotton for fiber use are consumed in their pure form. They are chiefly the low-cost ingredients in value-added food products or animal feed. Farm gate prices of most commodities are at historic lows in the United States, and the farmer's share of the consumer retail dollar spent on grains and vegetables is roughly four to seven cents (NASS 2001). Small improvements in farm productivity or reductions in input and labor costs are imperceptible to the final supermarket customer. Input traits such as pest and herbicide tolerance have no direct cost benefit to the end consumer of the engineered crops, and it is worth calculating whether any input trait in agriculture or in forestry could have a direct consumer price benefit.

Herbicide tolerance has the demonstrated environmental benefits of replacement of more toxic herbicides by chemicals such as glyphosate and the adoption of no-till farming methods that save labor, fuel, soil, and water. For the *Bt* crops, the demonstrated benefits are reduced dependence on more toxic pesticides and reduced costs of production (Carpenter and Gianessi 2001; Marra et al. 2002). Still, the public has not perceived the input traits of herbicide and pest resistance in food crops as directly benefiting it; the reasons for this hold lessons for tree biotechnology.

### Don't Ask, Don't Tell

The general public is unaware of—and may not want to know—the quantity and nature of chemicals used on crops in industrial agriculture and the negative impacts of modern farming. To say that genetically engineered herbicide tolerance allows the use of "less herbicide" draws attention to the use of chemicals and associates the consumer product with chemical intensive and "unnatural" farming. Although no-till farming is an important advance, complex environmental issues of destructive farming, non-point source water pollution, and soil loss are distant from consumers' decisions about food purchase. The dramatic growth of the market for organic foods is more a testament to fears about the safety of foods and perceptions of nutritional value than an indication of environmental concern. The creators of food brands want to associate food with a natural rather than a destructive image of farming, and less harm done is not compelling as an association with delivering more societal good. Second, all the data on the economic and environmental benefits of herbicide resistance in food crops have not been transparently shared by the sponsoring companies, and the data have been publicly questioned by critics of biotechnology.

Third, the chemicals that the input traits reduce or replace are produced and sold by the very same agrochemical companies, used by the same customers, and approved by the same regulatory authorities that sell, use, and approve the genetically engineered seeds. Drawing too much attention to the chemicals and their relative food and environmental safety (which should be comparable when used within approved limits) might also draw criticism to the agrochemical industry, the regulatory authorities, and the farmers—a no-win situation for everyone. The last reason why herbicide resistance was not marketable to consumers is that those who promoted its benefits are also those who would profit the most by selling the seed and the herbicide. Balanced, then, against no direct price benefit and unappreciated indirect benefits are the public's fears about risks to the environment and to human health: Why should I bear even remote or unknown risk, if others profit and I don't benefit? demands the wary consumer.

A similar set of reasons explains why consumers do not perceive that they are directly benefiting from pest resistance traits. The simple description of the *Bt* toxin crops is that the crop produces its own insecticide instead of requiring the use of insecticidal chemicals, and that draws attention to the use of chemicals and to the fact that the consumer may be eating a "toxin," albeit one harmless to humans. Moreover, the environmental benefits claimed for foods that contain biotech-derived ingredients were not straightforwardly communicated to the customers. Environmental activists, however, vociferously communicated charges of threats to nature and beneficial insects, as well as to the purity of organic crops. Drawing consumer attention to the EPA's and FDA's findings that both genetically engineered crops and chemical pesticides can be used safely is accurate but is not consoling to today's consumer. Perhaps another mistake in communicating the benefit of genetically engineered crops was the industry's resistance to product labeling in a consumer society where product advantages are prominently publicized on product labels and in broadcast media advertising. A citizen logically wonders, If this is so good for me and the environment, why isn't it advertised on the label?

## Broader System Impacts

Seed traits influence the cropping system and the relationship between farmers and seed suppliers. The benefits of herbicide resistance are only realized by farmers rich enough to afford herbicides. Like pesticidal crops, herbicide-resistant crops may only be available to farmers in countries with regulatory and intellectual property regimes that allow, and provide incentives for, bio-technological seed development. The high costs of product development and regulatory approval of new traits make the crops of modern monoculture the only viable commercial target for the private sector, and corporations depend on such cropping systems' remaining and even expanding. The same is likely to be true in forest biotechnology—that is, very large monocultural tree plantations may be the only structures to attain the scale, the economic ben-efits, and the environmental containment necessary for commercial viability. Thus, to the critics, genetic engineering of input traits in trees is seen as directly promoting monocultural tree plantations at the expense of a more biodiverse approach to forest management (Owusu 1999; Sampson and Lohmann 2000).

To a world that does not perceive trees as agricultural crops but as symbols of nature, trees that produce bacterial insecticides, or that are made to be sprayed with herbicides, are not likely to be accepted if there are perceived environmental risks. Both transgenic trees designed to be herbicide tolerant solely for the benefit of survival at the seedling stage, and plantations of trees expressing *Bt* toxin irrespective of pest levels seem to be poor starting points for the industry. The message of the agricultural crop experience for forest biotechnologists appears to be clear: Commercialize output and social utility traits well before commercializing the input traits that might increase chemi-cal use, promote chemical use, or draw attention to chemical use and the "unnaturalness" of tree plantations or genetically modified trees.

## Pressures and Choices Facing the Forest Industry

Discussion of forest biotechnology often starts with the question, Does the world need transgenic trees? This is an important question and may be an important guide for public sector research and development. However another, more immediate question might be, Does industry want transgenic trees? Agricultural biotechnology has shown that when there is a powerful economic motivation for industry, the genetically engineered products will be developed. Strong financial pressure on Monsanto from its huge invest-ments in seed companies and rising valuation on Wall Street was a powerful accelerator for GE crop introduction and created a competitive environment that demanded a similar response from the company's agrochemical rivals, including Dow, DuPont, Aventis, and Novartis (now Syngenta) (Enriquez and Goldberg 2000; Specter 2000). The creation of Arborgen in 1999, ini-tially a $60 million joint venture by Monsanto, Westvaco, International Paper, and Fletcher Challenge, reflects the corporate instinct to minimize the risk of this new field through collaboration and very modest investment. Monsanto's later withdrawal was consistent with its withdrawal from anything not associated with its core agricultural business. A risk reduction and learning strategy of the timber industry is its joint sponsorship of research, including the Tree Genetic Engineering Research Cooperative (TGERC) at Oregon State University (Doering and Parayre 2000; OSU 2002). Although it does not appear that there will be a forest industry leap into biotechnology, potential foreign and domestic competitors will likely force companies to begin to develop positions in the field in advance of either clear demand or mature science and product design.

A crude snapshot of the forest product industry (timber, pulp, and paper) shows a sector under regulatory pressure, facing rising global competition, and striving to achieve modernization, value-added products, and an improved reputation. Public pressure from activist organizations, on all parts of the value chain, from logging companies, to paper mills, to the big do-it-yourself stores, has been great. Resulting from that pressure are increasing demand for products from certified forests and an imperative for the industry to transform itself from one of the last extractive industries to its potential as a sustainable industry based on renewable resources (Gereffi et al. 2001; Doering et al. 2002). As sustainable forest management practices and plantation management improve, there may now be a strong financial argument for investment in GE tree development and regulatory approval.

## New Capabilities and Culture

This context is a contentious one for forest biotechnology. The vision of advanced, proprietary technology and genetically engineered, fast-growing, branchless, supertrees must be alluring to some leaders of an industry of bulldozers, chain saws, and pulp mills. Moving out of contested forests and into privately owned plantations might also be attractive. At the same time, product development and regulatory costs and the vision of anti-biotechnology protesters destroying test plots of trees or assailing company representatives must make the same business leaders distinctly queasy.

The development of genetically engineered trees calls for a significant cultural and technical change in the industry. For the agrochemical and pharmaceutical giants, the technology and regulatory processes of genetic engineering were not entirely new and played to their competitive strengths. But the molecular biology, ecological testing, compliance procedures, intellectual property concerns, and regulatory processes for commercializing a transgenic tree are not part of the traditional and current capabilities of the forest product industry.

The first transgenic tree plantations will require measures for biological and physical containment, intensive ecological monitoring protocols, and

**TABLE 8-1.** The Possible Pitfalls of Forest Biotech

— Lack of expertise that bridges sectoral gaps and interdisciplinary gaps
— Lack of analysis of global or local needs
— No supporting policy to protect forests and to justify plantations
— The seeking of public trust based on altruistic claims of distant environmental
  benefits
— Mistaken assumptions of supply–demand relationships
— Public rejection of tree plantations
— Failure to engage stakeholders in product and field trial design
— Failure to create public–private partnerships for product development and testing
— Commercial pressure to go to market too early
— Lack of regulatory, scientific, and stewardship capacity
— Pressures for unethical practices created by regulatory costs
— Imitation of agricultural products; i.e., the lure of easy input traits
— Failure to develop output traits with direct and demonstrable public benefits
— Trait and species choices that are science-driven rather than pulled by the market
— Overvaluation and pursuit of patents
— Long-term product liability, plantation monitoring, and stewardship issues

fences or barriers for economic and physical security. The long-term impacts of transgenic plants are unknown and will be trait-specific, and thus one can only speculate about the potential long-term liabilities. It is likely that financial resources, strategy, culture, and physical methods for long-term product stewardship are particularly important for transgenic trees. All these features will raise costs and demand skilled labor and new management methods. A company with a transgenic tree needs a new public acceptance of trees as crops grown on farms; new science and new regulatory systems; as well as new management practices for product development, product stewardship, and plantation management. Each of these demands lowers the probability that the companies at the front of the learning curve will be able to launch their new product profitably (Table 8-1).

## An Uncertain Role for Intellectual Property

A new forest biotechnology industry also faces difficult choices about intellectual property. Expensive to acquire and maintain, intellectual property rights are also highly controversial. Intellectual property ownership is central in the history and development of genetically engineered products. Patents are the foundation of the pharmaceutical industry and are credited with creating the conditions for the birth of the biotechnology industry. The gene patent race accompanying the effort to map the human genome reflects the continued importance of patents to the industry (Regaldo 2000). Intellectual property has also been one of the most contentious issues in the opposition to biotechnology, which has challenged the very validity of patenting life forms, the use of patents for economic control and competitive advantage, and the patenting of species considered to be in the public domain or to be part of the natural patrimony of developing countries (Altieri and Rosset 1999). Although it is commonly believed that patents are extremely valuable, there is also an analytic literature suggesting that patents are often overvalued, do not create strong competitive barriers, and have lower economic value and strategic utility than is often assumed (Mazzoleni and Nelson 1998; Cohen et al. 2000).

A tree biotechnology initiative will have to deal with the large suite of patents that already exist on molecular genetic methods and on the genes likely to be used to create a transgenic tree. But should tree technologists seek to patent engineered species and their underlying technology? If patents on genetically engineered trees are owned by the same companies that will grow and process the trees, do those companies need the same intellectual property protection as for seeds that may pass to other companies through a complex distribution chain? Are the costs of the patents, in direct terms and indirectly from potential societal opposition, justified when weighed against the extremely long life cycle of trees, the ease with which ownership may be established and protected, the rapid development of new technologies, and other means to protect property? The case for patenting trees is not obvious and merits analysis. Patenting may be a weak and incorrect strategic path for the forest biotechnology industry.

## The Roles of Science and Scientists

Besides the external pressures on the industry and the need for new capabilities, the scientific community itself can both facilitate and confound good decisionmaking by the industry. Biotechnology has been science driven, as people have looked for applications for new discoveries and ways to derive economic value from them. In areas of biology less well funded than human biomedicine, genetic engineering creates the possibility of attracting increased funding and greater scientific interest, as well as enriching scientists, investors, and research institutions through patents and the establishment of new biotechnology companies (Smith et al. 1999). For scientists who have spent lifetimes studying forest and tree biology, genetic engineering is a powerful tool to unlock scientific mysteries. Their enthusiasm for the science and technology is real and understandable. Whether the motivation of scientists is the purest interest in discovery, a genuine hope for sustainable technologies, or the desire for recognition and funding, there is a powerful confluence of reasons for scientists to be excited and to promote genetic engineering of trees. Food crop biotechnology has shown that social and environmental concerns fall outside the expertise and influence of scientists in both the public and private sectors. However, it is very much in the interest of those same scientists that their research and products be aligned with, and be responsive to, society's concerns, so as to ensure political and financial support of basic research.

The research scientists at the forefront of exploring genetic engineering of trees are quite logically the same ones companies seek as advisers and collaborators for exploring commercialization. Because this was also the case with crop biotechnology, why then was industry so totally unprepared to address and resolve the accompanying environmental and social issues? The reasons lie in the belief system of "sound science" and the absence of other scientific and social science viewpoints. The molecular biologists and the corporate strategists each thought that the other party comprehended and had under control the potential risks of the products. Missing from the implementation were ecologists and representatives of civil society who might have guided product design and introduction. Today, the agricultural biotechnology companies have broadened their scientific base and put in place high-level advisory boards with representation from diverse societal groups and organizations, though it is too early tell how those boards are affecting company actions. (The author is a member of Monsanto Corporation's Biotechnology Advisory Council.)

The mantra of "sound science" is repeated in debates on biotechnology with the implication that if only they understood the science, people would embrace the products of biotechnology. But sound science does not shape the marketplace and is low on the list of bases of consumer choice. Fears, desires, and price shape consumer acceptance, as is reflected in the cars we drive, the vitamins we take, the clothes we wear, the foods we eat, and the risks we bear for our pleasure and convenience. If the genetic engineering of forests is guided solely by "sound science" and scientists, the signals of the marketplace may well be missed. The pressures on industry and the motivations of scientists together create a force for the development of transgenic trees; that impetus can be harnessed for societal benefit only if other sectors of society become engaged to establish both goals and accountability mechanisms.

The scientific enterprise serves society's needs for new knowledge, new processes, and new technologies to improve the quality of our lives. The social rate of return on our considerable investment in basic research, such as that in agriculture, for example, are estimated to be very high (Alston et al. 1998). The scientific community is often called upon to produce "reliable knowledge" to resolve fears and uncertainties about natural risks and ones we create ourselves, such as assessments of asbestos risk, the impact of climate change, biological weapon threats, and questions of water safety (Gibbons 1999). There is a cost both to scientific progress and directly to taxpayers whenever a societal concern or a politically motivated fear requires scientists to do safety assessments and to take precautionary actions. The trust that society will grant to scientists and scientific institutions depends on its view of the legitimacy of the institution, the degree of democratic accountability, and the ethical acceptability of the base assumptions (ESRC 1999).

The environmental activists and commercial or political interests that oppose genetically engineered trees may increase short-term costs to society; yet they may also force the development of environmentally safe products that serve public needs and drive innovation. Controversies about agricultural biotechnology issues, such as the food and feed safety of *Bt* crops and their impacts on monarch butterflies, gene transfer to maize landraces in Mexico, and gene flow in canola, have furthered scientific knowledge, improved regulatory processes, and necessitated the invention of new methods for gene targeting, selectable markers, plant reproduction, gene mapping, ecological monitoring, and engineering of complex traits. The scientific revolution of molecular biology is recognized as the most powerful tool yet developed to improve our health and our lives, and with its power come both danger and responsibility. Genetically engineered plants are humanity's first self-replicating technology, and the costs of the unintended consequences are unknown. We know that there are social, economic, and environmental consequences—both good and bad—to genetically engineered crops, but we do not yet know how to design these plants to be fail-safe and safe-to-fail. Groups that fear technological change and distrust the companies and governments that control new technologies can increase commercial costs and change scientific plans to both society's benefit and its detriment. Those who have political, professional, and financial interests in accelerating forest biotechnology may similarly influence the scientific and regulatory agenda to both social risk and benefit. Such is the give-and-take of technology development in a transparent and democratic society, and it will be decades in the future before we can ascertain how the tension between precaution and promotion with regard to crop and tree biotechnology has served humanity.

## The Forest Products and Agricultural Biotech Industries: The Differences

In this chapter I have speculated on analogies between genetically engineered trees and genetically engineered crops, and my central thesis has been that the forest industry can learn from crop biotechnology's experience. There are, however, important differences between the two industries that may help prevent the repetition of some of the errors of crop biotechnology commercialization (Table 8-2).

The first difference is that in the case of GM trees, the technology developers are also the customers. The companies sponsoring research and development in genetically engineered trees are forest landowners and timber prod-

**TABLE 8-2.** Key Ways That the Fiber Industry Differs from the Food Biotechnology Industry

— The leading technology developers are the customers of the product.
— There is no agrochemical industry equivalent, with less financial pressure to introduce products.
— Limited intention exists to export and sell genetic stocks.
— Understanding of commercial needs originates in the industry.
— The scientists are tree and forest biologists with appreciation of forest systems.
— The public recognizes the need to reduce the ecological damage of the forest industry.
— A transition to plantation-based forestry is under way.
— Long planning time frames are usual in tree science and the timber business.
— Naturalness is not a consumer quality of paper and timber, as for food.
— The fiber system is simpler and more public than the food system.
— Fiber is less contentious than food.
— Forests and trees have higher symbolic value in the United States than does food.

uct companies, so that the understanding of industry needs that guides research and development originates in the industry itself. There seems to be no explicit interest in the export or sale of transgenic seedlings, though the high cost of product development will probably create pressure for exactly such a value-capture strategy. The molecular biologists involved in engineering trees were first trained as tree and forest biologists and so may be more likely than their crop science colleagues to consider the natural ecological context and the complex forest system in which GE trees may interact.

The forest industry is already undergoing significant change and modernization, and genetic engineering, rather than catalyzing disruptive change, as it did in the old chemical and drug companies, exists in the context of other trends toward modernization and greater environmental and social responsibility. Genetic engineering will not have to displace other approaches, and in fact, molecular genetics naturally complements and accelerates traditional tree breeding. It may take a while for people to see trees as crops, but the expansion of plantations is under way (FAO 2000). The tree industry does not have the long legacy of infrastructure, inputs, and planting practices that agriculture does; there may be some chance of a biodiverse and "ecosilvicultural" practice developing without the high barriers that exist to changing agriculture to a different model. A transgenic tree plantation may not seem much more unnatural than does the plantation itself, and genetically engineered trees may be acceptable in the context of a gradual and environmentally responsible transition from natural forests to tree plantations.

The negative ecological impacts of the timber and pulp industry are already a powerful narrative in the public mind. Because the forest product industry does not have the benign image of farming, it may be possible to show that plantations and engineered trees do "less harm" than logging in natural or public forests, and thus they may be viewed as progress. Many timber companies have recently engaged in public relations campaigns to position themselves as stewards of the forests, and they may consider GE trees as conflicting with that hard-fought message. However, "naturalness" is not a value that consumers associate with most timber and forest products; we do not seek a natural quality in our lumber or copy paper, as we do in a piece of corn or other vegetable. This is shown by the relatively minor objections leveled against genetically engineered cotton, in contrast to uproars over genetically engineered corn and wheat. Cotton is also the one crop for which there are the clearest data showing reduced use of harmful chemicals as a result of genetic modification (Carpenter and Gianessi 2001). The approval of *Bt* cotton planting by farmers in India and emerging data from small cotton farmers in China have provided the most compelling evidence of the promised "scale neutrality" of biotechnology; when all the GE inputs are in the seed, the small farmer benefits without major capital investment (Huang et al. 2002a). A consumer may fidget a moment to think that their jeans or underclothes contain cotton from genetically engineered plants. But learning that provokes a less visceral response than the discovery that their breakfast or lunch contains ingredients derived from genetically engineered crops.

The fiber system is simpler than the food system because transgenic trees may be developed by vertically integrated harvesters and processors. In many regions, there are fewer players, fewer products, fewer species and culture methods, and simpler value chains than in food supply. That simplicity may make it easier to design products and to develop value chain relationships where interests are better aligned than in the current path for food crops, in which genetically engineered seeds go from biotechnology company to farmer, to processors, to traders, to food companies, to supermarkets, and finally to consumers and restaurants.

Time may also be on the side of genetically engineered trees. The mindset of the forest product industry is one of much longer time frames than in the crop industry, which is based on annual cycles. It is standard practice for forest companies to think in 5, 10, and 20 year periods. The slow growth of trees ensures that there will be no fast product introduction; delays of a year to get the best transformant or to choose a proper genetic background will have less impact on the overall rate of GE tree commercialization. Tree scientists are also a patient lot. Importantly, too, there is no time or financial pressure on the industry from the public financial markets to meet or exploit the promises of the technology quickly.

Balanced against all these comparisons, which favor commercialization of genetically engineered trees, is the symbolic value of forests and trees. The American biotechnology companies did not account for the cultural symbol-

**TABLE 8-3.** Conditions for a GE Tree Market

— Social utility is the foremost concern.
— Design for the environment is of highest priority.
— GE trees reduce rather than create an environmental threat.
— Business motives and plans are transparent.
— Business communications and actions are aligned with investments.
— Stakeholders are engaged in decisionmaking.
— Private and public investment are balanced.
— The study of ecological impacts of transgenic tree plantations is well funded and precedes commercialization.
— Public policy and public opinion support tree plantations' link to forest conservation.
— Value capture and business strategy are not based in patents.
— The first developed traits are output traits.
— Stakeholders that perceive risks can make choices and perceive benefits.
— Target markets have appropriate regulatory capacity.
— Region-specific products are developed.
— Technology is applied to serve needy populations and to protect native forests.

ism and importance of agriculture in Europe, and biotechnology activists underestimated the desires of developing countries to use biotechnology and to make their own technological choices. The typical industry-sponsored public opinion survey asks about possible acceptance of benefits and not, Would you like a plantation of genetically engineered trees in your backyard? Timber product companies may be in for a shock regarding how the public feels about its trees, especially if the genetically engineered tree does not directly connect to the protection and renewal of forests. Barring significant financial benefit and undisputed environmental benefit, society may simply not tolerate "Frankentrees." Still, the framework conditions for a market for transgenic trees are conceivable when abstracted from our experience with genetically engineered crops (Table 8-3).

## Creating a Sustainable Future

I have never seen a deep analysis of the role of tree biotechnology that considers the values that forests and trees deliver in the context of specific nations, social values, regulatory structures, public policy, resource flows, timber substitution, and economic scenarios. But we cannot analyze the utility of genetically engineered trees without seriously asking, Compared to what? An opportunity exists to develop sustainable silviculture that is part of integrated management of productive forests, working landscapes, and protected forests to maximize ecosystem goods and services for all human uses.

Tree plantations do not have crop agriculture's legacy of monoculture or an entrenched system of production and production inputs, and they might still be designed for biodiversity.

The world's most biodiverse forests and last frontier forests are threatened by development, conversion to agricultural lands, mineral and oil extraction, fragmentation by roads, overhunting and overlogging, global climate change, and the destruction of water resources. Billions of people in the world have pressing needs for the tree-derived energy, paper, and materials that are important components of economic development, improved health, literacy, and commerce. And forests are central in protecting watersheds, purifying air and water, stabilizing climate, and protecting species diversity; they allow people to enjoy cultural and spiritual values, tourism, and recreation. In a world whose population will grow by a third in the next 25 years (chiefly in less developed countries), imaginative solutions, with a place for technology, are essential to meet global needs for water, materials, energy, and paper. They will require keeping sight of the sustainable forestry goals beyond the transgenic trees.

During the next 5 to 10 years, the public and private sectors will make critical decisions about investment in genetically engineered trees. Risk or benefit is not intrinsic to genetic engineering. Experience with the genetic engineering of crops has shown that cultural, environmental, and economic risks and benefits from genetic engineering are achievable. The risk and the benefit of a genetically engineered plant stem from the expressed trait, how it fits into a culture system, who benefits, and the risk of gene flow. Tree biotechnology has not yet crossed the proof-of-concept threshold for either risk or benefit, and what traits and species will be chosen for commercial modification remain uncertain. We can make wise choices, as citizens, scientists, and businesspeople, about how to develop the technology, with what safeguards, and to what ends. How can genetic engineering equitably and safely serve the needs of sustainable development? Our thoughtful, creative, and rigorous consideration of this question should be limited only by our knowledge at the moment. We must not lack the imagination and courage to envision and realize creative research, fair processes, shared benefits, and a sustainable future.

# References

Alston, J.M., G. Norton, and P.G. Pardey. 1998. *Science under Scarcity: Principles and Practice for Agricultural Research and Priority Setting.* Wallingford, UK: CAB International.

Altieri, M.A., and P. Rosset. 1999. *Ten Reasons Why Biotechnology Will Not Ensure Food Security, Protect the Environment and Reduce Poverty in the Developing World.*

Berkeley, CA: University of California, Berkeley, and Institute for Food and Development Policy, Oakland, CA.

Arnold, M.B., and R.M. Day. 1998. *The Next Bottom Line: Making Sustainable Development Tangible.* Washington, DC: World Resources Institute.

BIO (Biotechnology Industry Organization). 2002. Plant-Made Pharmaceuticals. http://www.bio.org/pmp/ (accessed January 1, 2003).

Carpenter, J.E., and L.P. Gianessi. 2001. *Agricultural Biotechnology: Updated Benefit Estimates.* Washington, DC: National Center for Food and Agricultural Policy.

CBI (Council for Biotechnology Information). 2002. *Growing More Food: How Biotechnology in a Seed Can Enhance Global Food Security.* Washington, DC: Council for Biotechnology Information.

Challener, C. 2001. A Changing Landscape for Agrochemicals. *Chemical Market Reporter* 260(23): 9–12.

Cohen, W.M., R.R. Nelson, and J.P. Walsh. 2000. *Protecting Their Intellectual Assets: Appropriability Conditions and Why U.S. Manufacturing Firms Patent (or Not).* NBER W7552. Cambridge, MA: National Bureau of Economic Research.

Coile, Z. 2002. Bush Speeds Logging Plan. *San Francisco Chronicle,* Dec. 12, 3.

Daar, A.S., D.K. Martin, S. Nast, A.C. Smith, P.A. Singer, and H. Thorsteinsdottir. 2002. *Top 10 Biotechnologies for Improving Health in Developing Countries.* Toronto: University of Toronto Joint Centre for Bioethics.

Doering, D.S., A. Cassara, C. Layke, J. Ranganathan, C. Revenga, D. Tunstall, and W. Vanasselt. 2002. *Tomorrow's Markets: Global Trends and Their Implications for Business.* Washington, DC: World Resources Institute, United Nations Environment Programme, World Business Council for Sustainable Development.

Doering, D.S., and R. Parayre. 2000. Identification and Assessment of Emerging Technologies. In *Wharton on Managing Emerging Technologies,* edited by G.S. Day and P.J.H. Schoemaker. New York: John Wiley, 75–98.

Elkington, J. 1997. *Cannibals with Forks.* Oxford, UK: Capstone Publishing.

Enriquez, J., and R.A. Goldberg. 2000. Transforming Life, Transforming Business: The Life-Science Revolution. *Harvard Business Review,* March–April, 96–104.

ESRC (Economic and Social Research Council). 1999. *The Politics of GM Food: Risk, Science and Public Trust.* Special Briefing No. 5. University of Sussex, UK.

EU (European Union). 2002. Genetically Modified Organisms (GMOs). http://europa.eu.int/comm/food/fs/gmo/gmo_index_en.html (accessed December 18, 2002).

FAO (UN Food and Agriculture Organization). 2000. *Global Forest Resources Assessment.* Rome, Italy: FAO.

Fenning, T.M., and J. Gershenzon. 2002. Where Will the Wood Come From? Plantation Forests and the Role of Biotechnology. *Trends in Biotechnology* 20: 291–296.

Freese, B. 2002. *Manufacturing Drugs and Chemicals in Crops: Biopharming Poses New Threats to Consumers, Farmers, Food Companies and the Environment.* Washington, DC: Friends of the Earth.

Gaskell, G., N. Allum, M. Bauer, J. Durant, A. Allansdottir, H. Bonfadelli, D. Boy, et al. 2000. Biotechnology and the European Public. *Nature Biotechnology* 18: 935–938.

Gereffi, G., R. Garcia-Johnson, and E. Sasser. 2001. The NGO–Industrial Complex. *Foreign Policy,* July–August, 56–65.

Gibbons, M. 1999. Science's New Social Contract with Society. *Nature* 402(Supp.): C81–C84.

Huang, J., R. Hu, S. Rozelle, F. Qiao, and C.E. Pray. 2002a. Transgenic Varieties and Productivity of Smallholder Cotton Farmers in China. *Australian Journal of Agricultural and Resources Economics* 46(3): 367–387.

Huang, J., C. Pray, and S. Rozell. 2002b. Enhancing Crops to Feed the Poor. *Nature* 418: 678–684.

Juma, C. 1999. *Science, Technology and Economic Growth: Africa's Biopolicy Agenda in the 21st Century.* Addis Ababa, Ethiopia: UNU/INRA Distinguished Annual Lectures.

Lenzner, R., and T. Kellner. 2000. Corporate Saboteurs: They Wrecked Monsanto, Now They're after the U.S. Drug Industry. Is Your Company Next? *Forbes,* November 27, 156–168.

Mann, C.C., and M.L. Plummer. 2002. Forest Biotech Edges Out of the Lab. *Science* 295: 1626–1629.

Marra, M.C., P.G. Pardey, and J.M. Alston. 2002. *The Payoffs to Agricultural Biotechnology: An Assessment of the Evidence.* Washington, DC: International Food Policy Research Institute.

Matthews, E., R. Payne, M. Rohweder, and S. Murray. 2000. *Pilot Analysis of Global Ecosystems: Forest Ecosystems.* Washington, DC: World Resources Institute.

Mazzoleni, R., and R.R. Nelson. 1998. Economic Theories about the Benefits and Costs of Patents. *Journal of Economic Issues* 32: 1031–1052.

Monsanto (Monsanto Corporation). 2002. Our Commitments: Environmental & Social Responsibility: Monsanto Pledge. http://www.monsanto.com/monsanto/layout/our_commitments/default.asp (accessed December 31, 2002).

Moore, J. 2001. More Than a Food Fight. *Issues in Science and Technology* 17(4): 31–36.

NASS (National Agricultural Statistics Service). 2001. Agricultural Prices. http://www.usda.gov/nass/nasshome.htm (accessed March 1, 2002).

NCFAP (National Center for Food and Agricultural Policy). 2002. *Plant Biotechnology: Current and Potential Impact for Improving Pest Management in U.S. Agriculture.* Washington, DC: National Center for Food and Agricultural Policy.

NSB (National Science Board). 2002. *Science and Engineering Indicators—2002.* Arlington, VA: National Science Foundation.

OSU (Oregon State University). 2002. Tree Genetic Engineering Research Cooperative (TGERC). http://www.fsl.orst.edu/tgerc/ (accessed December 15, 2002).

Owusu, R. 1999. *GM Technology in the Forest Sector: A Scoping Study for WWF.* Washington, DC: World Wildlife Fund.

Peña, L., and A. Sequin. 2001. Recent Advances in the Genetic Transformation of Trees. *Trends in Biotechnology* 19(12): 500–506.

Pew (Pew Initiative on Food and Biotechnology). 2002. Environmental Savior or Saboteur: Debating the Impacts of Genetic Engineering. http://pewagbiotech.org/research/survey1-02.pdf (accessed December 18, 2002).

Pollack, A. 2002. Spread of Gene-Altered Pharmaceutical Corn Spurs $3 Million Fine. *New York Times,* Dec. 7.

Priest, S.H. 2000. U.S. Public Opinion Divided over Biotechnology? *Nature Biotechnology* 18(September): 939–942.

Regaldo, A. 2000. The Great Gene Grab. *Technology Review,* September–October, 49–67.

Repetto, R., and D. Austin. 2000. *Pure Profit: The Financial Implications of Environmental Performance.* Washington, DC: World Resources Institute.

Sampson, V., and L. Lohmann. 2000. *Briefing 21. Genetic Dialectic: The Biological Politics of Genetically Modified Trees.* London, UK: Corner House.

Seuss, D. 1971. *The Lorax.* New York: Random House.

Smith, K.R., N. Ballenger, K. Day-Rubenstein, P. Heisey, and C. Klotz-Ingram. 1999. *Biotechnology Research: Weighing the Options for a New Public–Private Balance.* Washington, DC: U.S. Department of Agriculture.

Specter, M. 2000. The Pharmageddon Riddle. *The New Yorker,* April 10, 58–71.

Tischner, U., and M. Charter. 2001. *Sustainable Product Design. Sustainable Solutions: Developing Products and Services for the Future.* Trowbridge, UK: Greenleaf Publishing.

Victor, D.G., and C.F. Runge. 2002. *Sustaining a Revolution: A Policy Strategy for Crop Engineering.* New York: Council on Foreign Relations.

Wambugu, F. 1999. Why Africa Needs Agricultural Biotech. *Nature* 400, July 1. 15–16.

Weil, A. 2001. The Future of Transgenic Plants in Developing Countries. *Cellular and Molecular Biology* 47(8): 1343–51.

WRI (World Resources Institute, United Nations Development Programme, United Nations Environment Programme, and World Bank). 2000. *World Resources 2000–2001: People and Ecosystems; the Fraying Web of Life.* Washington, DC: World Resources Institute.

# 9

# Environmental and Social Aspects of the Intensive Plantation/Reserve Debate

Sharon T. Friedman and Susan Charnley

O ur purpose in this chapter is to address the argument that increasing industrial wood production on plantations to meet future wood demand will reduce harvest pressure on existing old-growth and "natural" forests and lead to their protection. The work of Victor and Ausubel illustrates this argument: "[T]he main benefit of the new approach to forests will not reside within the planted woods, however. It will lie elsewhere, in the trees spared by more efficient forestry. An industry that draws from planted forests rather than cutting from the wild will disturb only one-fifth or less of the area for the same volume of wood. Instead of logging half the world's forests, humanity can leave almost 90 percent of them minimally disturbed. And nearly all new tree plantations are established on abandoned croplands which are already abundant and accessible" (2000, *138*). Other natural resource scientists have made this argument as well (e.g., Binkley 1997; Mather 1990; Pandey 1995; Sedjo and Botkin 1997; Siry et al. 2001; Whitmore 1999; Sedjo 2001).

These researchers view industrial roundwood production as an industry that should maximize its efficiency by using as little land and producing as much wood as possible. Industrial roundwood production on intensively managed plantations is assumed to be preferable to extensively managed production on natural forests. We, on the other hand, emphasize choice of silvicultural practice as appropriate in specific contexts. We ask what practices, including genetic engineering of forest trees, are best to use on a particular piece of land to meet a specific set of economic, social, and environmental objectives, with control over those decisions resting with landowners, forestry practitioners, local communities, and governments.

This chapter challenges the argument that intensive forest management practices on plantations prevent the degradation of natural forests, thereby protecting the ecological and social values of natural forests and making plantations environmentally and socially desirable. We question not only the ecological assumptions associated with this argument but the social assumptions as well. Both the ecological and the social effects of industrial roundwood plantations must be examined to fully understand their implications and to judge whether and how they will be beneficial for environmental conservation.

## Definitions: Plantations and Intensive Management

There is no simple dichotomy between "plantation forests" and "natural forests." The Society of American Foresters *Dictionary of Forestry* (Helms 1998) defines a plantation as a "stand composed primarily of trees established by planting or artificial seeding." There are three footnotes associated with this definition.

- A plantation may have tree or understory components that have resulted from natural regeneration.
- Depending on management objectives, a plantation may have pure or mixed species, be treated to have uniform or diverse structure and age classes, and have wildlife species commensurate with its state of development and structure.
- Plantations may be grown on short rotations for biomass energy or fiber production, on rotations of varying length for timber production, or indefinitely for other values.

By this definition, "plantation" simply refers to the origin of some of the trees in the stand. A plantation can be a mixed-species grouping for agroforestry, an enrichment planting of a rare species, or an intensively managed plantation for industrial wood production. Forest management and silvicultural systems exist on a continuum, and "natural" and "artificial" forests occupy different but overlapping points in a spectrum, making it sometimes difficult to distinguish natural forests and forest plantations for the purpose of generating worldwide statistics (Brown 2000).

Plantations may be established for a number of reasons:

- to compensate for a lack of resources from natural forests due to deforestation;
- to meet demand for lumber, pulp, and paper products (for domestic or export markets);
- to meet demand for high-quality species (teak, mahogany, Spanish cedar);

- to develop export markets;
- to provide domestic uses, such as firewood, posts, and home fences;
- to restore degraded sites and protect watersheds;
- to protect the genetic diversity of forest species ex situ;
- to provide forest uses on previously unproductive sites when natural forests are not accessible;
- to produce desired species that have failed to regenerate naturally; and
- to supply potential markets for carbon sequestration (Evans 1992; Whitmore 1999).

"Intensively managed plantations," as described in the literature, are most often managed with the objective of obtaining a high volume per hectare of one or a few uniform products, typically industrial roundwood or high-value hardwoods. However, plantations can also be intensively managed for products such as cork, latex, or nuts. It is important to note that there is no general agreement on what the term "intensive management" means. It is a value judgment and can refer to an array of silvicultural practices, each of which has a range of intensities and can occur on different temporal and spatial scales. Every stand planted is not necessarily managed intensively later; plantations can be extensively managed, and naturally regenerated stands can be intensively managed. The proposed intensity of management is a function of the silvicultural choices available, their economic and biological feasibility, consideration of potential impacts, and the landowners' objectives. "Extensive," "moderate," and "intensive" practices only have meaning in the context of specific sets of silvicultural prescriptions arrayed on specific landscapes.

The argument that intensively managed plantations will relieve pressure on natural forests and provide environmental and social benefits pertains to intensively managed industrial forest plantations that produce wood and fiber products. We therefore focus here on intensively managed plantations for industrial roundwood production (IMPIRs), while recognizing that multiple definitions and contexts are associated with the term "intensively managed plantations," that forests exist on a continuum, and that "intensive" is in the eye of the beholder.[1] First we address arguments relating to the environmental effects of IMPIRs. Then we turn to discuss the social effects of IMPIRs.

## Environmental Considerations

Industrial roundwood plantations are assumed to be beneficial because they provide cheap wood to distant people and jobs for local people. They are also assumed to relieve logging pressure on native forests, leaving them for biodiversity habitat, watershed protection, and other nonindustrial uses (Sedjo

2001; Victor and Ausubel 2001; Salwasser 2001). As Sedjo has put it, "An environmental implication of the increased productivity of planted forests due to biotechnology is likely to be that large areas of natural forest might be free from pressures to produce industrial wood, thereby being better able to provide biodiversity habitat" (2001, *29*). Although that may be true in some cases, industrial roundwood production may not be the only, or even the most important, reason for pressure on many forests; pressure from sources other than commercial logging also has to be considered.

There are two broad categories of stressors on forests: those causing changes in forest area and those causing changes in forest structure and species composition.[2]

## Changes in Forest Area

The idea that industrial forest plantations will relieve pressure on forests suggests that logging is an important contributor to the degeneration of forests. Although that may be true in the sense that the removal of most of the overstory fundamentally changes the composition, structure, and function of a forest for some time period, other activities can also lead to the removal of forest overstories and do so more permanently than logging does. Most forest loss is caused by the conversion of forests to agricultural and residential development (Salwasser 2001). Although slash-and-burn agricultural fields under shifting cultivation are intended to revert periodically to forest, other kinds of agricultural practices, and high demand for agricultural land, may preclude forest regeneration. The conversion of forest to urban land uses is even more likely to be permanent. Establishing IMPIRs will not relieve these pressures.

The amount of pressure on forests for land use change is expected to vary from country to country, and from region to region within a country, depending on the demand for, and economics of, alternative land uses. The southeastern United States is a major timber-producing region of the world, producing 15.8% of the world's timber in 1997, as described in the Southeastern Forest Assessment (Wear and Greis 2002). The assessment also points to urbanization as the major threat to forest land use in that part of the United States, with trade-offs among urbanization, agricultural development, and forest uses dependent on assumptions about future prices for agricultural and forest products. According to the assessment, land goes out of agriculture and into forestry when forest products are more profitable than agricultural products (these trends are also sensitive to local pressures for residential development). Thus higher prices for wood products may help to keep some lands in forests and out of agriculture. If cheap fiber from intensively managed plantations were to lower prices for wood products, logging in natural forests may become unprofitable, with the result that land is less likely to be in forests and more likely to be in agriculture. Sargent and Bass (1992) make the same argument.

Lower prices for a commodity can also increase demand and convert the commodity from high-value to low-value uses. Wood in particular has many low-grade markets in pulp and processed board applications. Increased supply and lower prices can increase the use of timber for low-value purposes, increasing deforestation. They can also favor the low-cost exploiter of the forest, who may not be the best environmental steward (Doering 2001).

Hays (2001) points out the importance of ownership for the fate of natural forests. Plantations can relieve logging pressure on native forests and lead to their protection if the forests are on publicly owned lands and if the social, political, and economic contexts are favorable. In New Zealand, for example, legislation passed in 1985 placed all native forest occurring on state-owned lands under the management of the Department of Conservation, which took a strictly preservationist approach, prohibiting any form of wood extraction (Maclaren 2001). Existing plantations on state-owned lands were made corporate and eventually privatized, with the idea that industrial wood production should take place on IMPIRs, using exotic species, and natural forests should be preserved on public lands (industrial wood production from extensively managed natural forests is not commercially competitive in New Zealand).

In the United States, most industrial roundwood production (91% in 1997) occurs on privately owned lands (Haynes 2003), and two-thirds of this production comes from timberlands owned by nonindustrial landowners. Only 19% of this roundwood is produced from intensively managed stands or plantations (which tend to be on industry-owned lands). The remaining 81% of industrial roundwood production on private lands in the United States takes place from natural forests under some form of custodial management. If wood production becomes unprofitable for private landowners because the increased efficiency of IMPIRs depresses prices, they may sell the forestland they own or convert it to other uses that are profitable, such as agriculture or residential development. For the public to ensure that current forestlands are maintained in forests, it may need to acquire land, create new tax incentives, or impose regulations. The marketplace can also reward forest preservation through mechanisms such as forest certification programs and conservation easements.

## Changes in Forest Structure and Species Composition

To achieve sustainable forest management, forest use must be balanced with forest conservation. If demand for commodity industrial roundwood production in natural forests were to diminish as a result of production from intensive plantations, demands on natural forests for other uses, such as wood harvesting for domestic purposes, the collection of nontimber forest products, and recreation, would remain.

**Pressures from Human Use.** Human pressure on forests comes from commercial, subsistence, and personal uses, both legal and illegal. For example, FAO estimated that just over half of total world roundwood production in 1999 was for wood fuel. Global production of industrial roundwood, on the other hand, has been dominated by developed countries, which produced 79% of the total (FAO 2000). This would suggest that to protect forests in developing countries, it is essential to find alternative fuels, whereas to protect forests in developed countries, it is important to produce more industrial roundwood. Rather than intensify production, however, developed countries tend to use certification and other, less-intensive forms of management (and Forest Stewardship Council-affiliated forest certification schemes do not permit genetically engineered trees of any kind).

As the FAO estimates show, pressure on forests for fuelwood is particularly high in developing countries. In India, for example, fuelwood was estimated to account for almost 90% of the total wood consumed in 1985, whereas only 7% was for timber and only 3% for pulpwood (Adkoli 1992). A better way to relieve the pressure on forests from demand for fuelwood might be to fund research on, or local purchases of, new technologies to provide heat for cooking—such as biogas or more efficient stoves—instead of establishing IMPIRs. On the other hand, plantations designed for fuelwood rather than industrial roundwood production can and do supply fuel locally and thus can remove pressure from natural forests (South 1999). Kohlin and Parks (2001) found that this held true in one area of India, provided that the plantation was located closer to the community than natural forest, making it more easily accessible than fuelwood from natural forests. Unless secondary environmental impacts would result from a given strain of tree (e.g., excessive water use), genetically engineered, low-cost strains of exotic trees may well be easily accepted for fuelwood plantings. For promoters of genetically engineered trees, this would be a relatively noncontroversial application, particularly if exotic species were used, so as to minimize concern about gene flow to native species.

There are separate global and local markets for a variety of high-value hardwood and softwood sawtimber and logs. Some species of particularly high value, such as teak, mahogany, and walnut, are currently grown in intensively managed plantations. Other species, such as redwood, Alaska yellow cedar, and other softwoods having a high value can be grown in plantations but are not currently produced that way for a variety of reasons (e.g., redwoods reproduce by sprouting in their natural range). Through genetic engineering, different species may be able to be grown economically in a plantation environment. The markets are complex, and there are some substitutions among species, but changes in supplies of these high-value products will not affect the global commodity-grade market for industrial roundwood. Conversely, IMPIRs of other species will not relieve pressure for high-value products. On the other hand, plantations for high-value wood products are likely to engender some of the same environmental and social concerns as developing them for lower value wood products.

Trees produce a variety of useful products apart from industrial round-wood. For example, leaves and branches provide animal forage; people, live-stock, and wildlife eat fruits and nuts; pollen is used for cosmetics and medicinals; cones, branches, and bark are used for making decorations; roots and bark are used for fiber; and syrup and honey production are associated with some species. Many other forest plants are also highly valued for food (e.g., berries, mushrooms, edible roots), decoratives, and medicine (e.g., gin-seng). As these products gain value in the global marketplace, and as people migrate and settle in different parts of the world, the amounts collected for both market and personal use increase. In addition, forest birds and mam-mals may be hunted for subsistence or for commercial purposes, providing income to people in forest communities. People also engage in noncommod-ity uses of forests, such as recreation and spiritual nourishment. These uses may be personal or commercialized through tourism.

All of these nontimber forest uses have ecological impact. Whether their impact is beneficial or detrimental to forests and their flora, fauna, and aquatic ecosystems depends on the level and manner of use. In India, for example, a major threat to natural forests has been the lack of control over access to them, resulting in overexploitation after logging (Ravindranath et al. 1996; Poffen-berger and McGean 1996). Grazing, cutting, and fires have made it difficult for natural forests to regenerate naturally, and so plantations of exotic timber spe-cies were established. However, the plantations did not meet local people's needs, and so they did not solve the problem of protection for natural forests. Lack of control over access to plantations has also become a problem. If natural forests (and plantations) are to be protected in India, controls on access and use must be put in place (Poffenberger and McGean 1996).

Intensive forest management focuses on growing industrial commodity-grade roundwood quickly and not on providing other timber and nontimber forest products and uses. IMPIRs will not help to reduce the pressures on forests that result from demand for fuelwood, high-value and specialty hard-wood and softwood timber products, and nonwood uses of trees and their associated flora and fauna. In fact, where IMPIRs are established on existing natural forestlands, the area of native forest available to produce those other benefits will decrease. Even if all IMPIRs were established on former agricul-tural land, the only value they would provide would be commodity-grade wood and fiber, and that would be the only pressure on forests that they might relieve.

**Pressures from Other Sources.** Sources of pressures on forest structure and species composition other than direct human use include climate change, pollution, drought, and the increasing movement of people and products, leading to rising threats from exotic pathogens and insects. IMPIRS are unlikely to have a direct impact on these.

Secondary changes in forest structure and composition may also occur as a result of forest management policies. For example, fire exclusion causes savannahs to become forested and changes the composition of forests in ways that promote the development of crowded, fire-prone stands. IMPIRs are unlikely to affect such changes other than by potentially lowering the price of industrial roundwood, making the thinning of fire-prone stands even less economical than it is today.

Even if all industrial roundwood demand were removed from forests, it is unclear that they would then be "saved" or "protected" because of the many other pressures discussed above. Again, some argue that establishing IMPIRs would have the opposite effect through lowering the price of wood and wood products. But high prices for forest products are an important protection against land conversion to higher value uses, such as agriculture or industry (Oliver 1999). Perhaps the most environmentally sound approach is to maintain the production of goods and services on natural forests to the extent possible and allow people to make money from sustainable forest uses; this is the approach that forest certification takes. The case of Menominee forestry in the United States provides an example (Davis 2000).

## Are IMPIRs More Environmentally Desirable Than Alternative Land Uses?

One assumption of the "protection of native forests" argument for IMPIRs is that native or extensively managed forests that are currently providing a wide variety of goods and environmental services will not be converted to industrial plantations; only previously "marginal" or "degraded" land will be used. In general, it is desirable to improve the conditions of these marginal lands by protecting soil and water quality and to use them, if possible, to produce some combination of products that can contribute to the well-being of local people. Certainly, industrial forest plantations could be one answer. However, depending on local needs and access to transportation and markets, a variety of equally or more desirable land use alternatives exists, including agriculture, agroforestry, or some mix of production systems that also can restore degraded lands. Scientists engaged in the genetic engineering of livestock and food crops might argue that these same marginal lands should be used to produce food, for subsistence or markets. Given the many possibilities for the genetic engineering of plants, it should not be assumed that commodity production of industrial roundwood for export is the best answer to the question of what should be done with marginal lands to balance environmental protection, human subsistence, and export needs. Furthermore, because the current ownership and location of these marginal acres are unknown, it is difficult to project how much of such land would ultimately best be used for intensively managed industrial roundwood production, even if potential alternatives and their benefits and risks were known.

Over the last few decades, some researchers have observed that most plantations have been established in areas previously used for agriculture (Victor and Ausubel 2000). That is not always the case, however. The scarcity of suitable land for new planting is the most common physical constraint on forest plantation development (Brown 2000). The physical terrain may be unsuited to forest plantations, or more frequently, the remaining arable land is more valuable in alternative uses, such as agriculture. Competition for land for agricultural purposes is greatest in developing countries with high population densities. For example, in Madagascar slash-and-burn cultivation currently is the principal threat to forests (Kremen et al. 2000). Where demand for arable land for agriculture is high, plantations may be more likely to be established in areas currently containing natural forests. Natural forest may also be cut and replanted with exotic species because the shorter cutting rotation for these species increases their potential for generating revenue. This practice has been widespread in Chile since the mid-1970s (Aagesen 2001). Alternatively, plantations may be established on state-owned forestlands that the government leases to industry to generate revenue and subsidize industry. In India, this practice has led to widespread clear-felling of natural forests and their replacement with monocultures of fast-growing species, such as eucalyptus (Hiremath and Dandavatimath 1996). In these cases, rather than contributing to the protection of native forests, plantations replace them, with dramatic consequences for biodiversity.

Even where plantations are established on marginal land previously used for agriculture, their ecological benefits may not outweigh their ecological costs. Victor and Ausubel (2000) estimate that worldwide, roughly half of the land developed for agriculture would revert to forest if left alone. The other half would likely revert to some other form of native habitat given time, be it grassland, savannah, or chaparral, which also has ecological value. If IMPIRs are established on old agricultural land instead of letting it revert to natural habitat (assuming that no alternative land use pressures exist and that natural habitat is an economically competitive and socially desirable land use), the biodiversity value of native habitat that might otherwise return will be forgone. To assume that there are no negative ecological consequences associated with establishing IMPIRs in areas previously used for agriculture because it requires no additional conversion of native habitat is to overlook the ecological value of allowing agricultural land to revert to its natural state over time. Land containing native ecosystems—be they forest, grassland, or shrubland—has high biodiversity and other ecological values, as well as social values such as game and food and medicinal plants, relative to IMPIRs.

Where IMPIRs are successful in relieving logging pressure on natural forests and those forests are protected and managed to provide ecological values and environmental services, the environmental effects of plantations themselves must be considered.

For each plantation, managers can choose from a suite of silvicultural practices, depending on the forest type and management objectives, as well as site-specific resources and conditions. Costs, expected gains in value, and social and environmental acceptability also enter into the choice of silvicultural treatments. Site preparation activities can include soil treatments such as bedding, creating drainage, and fertilization. Regeneration is the next phase and can include planting, seeding, or natural regeneration. Planting and seeding can be done with natives or exotics and with traditionally bred or (in the future) genetically engineered trees. After the trees are planted, there may be vegetation control using mechanical or chemical methods to reduce fuel, create open spaces for tree establishment, discourage vegetation that competes with trees, and provide protection from insects, disease, and animal damage. There may also be fertilizer treatments and possibly pruning throughout the rotation. Harvest methods can include clear-cutting or leaving different species, sizes, and numbers of trees through even-aged and uneven-aged silvicultural systems (Lautenschlager 2000).

Some of the silvicultural treatments described above have greater environmental impact and are more controversial than others, at least as observed in Canada (Lautenschlager 2000). For example, whereas thinning and pruning are expected to generate little concern from the public, chemical applications, particularly chemicals aerially applied, generate more concern and pose a greater environmental threat (Lautenschlager 2000; Spencer and Jellinek 1995). Chemicals, herbicides, and pesticides used on plantations can enter local watercourses and pollute the drinking water of local communities, especially in developing countries (Colchester 2000). Plantation establishment may also alter hydrological cycles and affect fish populations. When plantations contain large stands of exotic and/or transgenic trees, there is the risk of their escape and establishment in native forests. Where plantations contain genetically engineered trees, environmental concerns revolve around whether there might be genetic transfer to nearby wild populations of the same or similar species (Sedjo 2001) and the biodiversity impacts on other organisms in the food web (Hays 2001). All of this raises the question of whether sustainable wood production from natural forests or wood production on IMPIRs is the most environmentally sound choice.

Finally, some scientists have questioned the long-term ecological sustainability of IMPIRs. Short harvest rotations can cause loss of soil fertility and productivity (Bowyer 2001). Eucalyptus plantations have been found to cause soil erosion because of the lack of an understory and the burning of leaf litter to prevent wildfires (Bandaratillake 1996). The cultivation of monocultures may also increase the risk of catastrophic disease and insect infestations or increase the need to protect against them through use of chemicals or biocontrol (Bowyer 2001).

In some cases, IMPIRs may be successful in relieving pressure on natural forests and leading to their protection (see Maclaren 2001). Although there are environmental benefits associated with this outcome, what are the social consequences? In the next section, we examine the human dimensions of the plantation/reserve debate.

## Social Considerations

One of the difficulties of the industrial plantation/reserve debate is that it is not clear what role local people and their uses of land and resources are to play in either the plantations or the reserves. When the focus is exclusively on maximizing timber yields from plantations, other, previous uses of those acres (be they forest or nonforest) inevitably are squeezed out and with them, people's resource entitlements. Hence one set of concerns revolves around the reduction of alternative land and forest uses once the management focus is on maximizing timber yields from a given area. Associated with this process are shifts in property rights. A second set of concerns revolves around the establishment of reserves for forest protection, as a corollary of the plantation/reserve argument, and how that affects the use rights and activities of local people. A third set of concerns relates to the role of local people in decisions regarding how local forests, whether plantations or forest reserves, are to be used and managed.

### Impacts of Plantations on Local People

In contrast to plantations established to promote social forestry, which aim to meet the needs of local residents, IMPIRs are designed to meet national goals and priorities and to supply the global market in industrial roundwood. There may be some local benefits from IMPIRs, such as employment opportunities and the availability of residues and byproducts after felling for fuelwood or timber (Adkoli 1992). Indirect benefits include the reinvestment of revenues derived from plantation forestry locally in education, medicine, and infrastructure development, where it occurs (Tapp 1996). However, the literature documenting the social impacts of IMPIR establishment on local people suggests that IMPIRS are largely negative because they do not take local needs and concerns into account. Two major impacts are apparent: (a) the disenfranchisement of local people, who lose access to land and forest products for a variety of uses, undermining their ability to subsist; and (b) the outright displacement of local people from rural lands. Both result in increased migration, often to urban areas, by people who have become "environmental refugees" (people forced to leave their homes when ecological conditions make it impossible for them to remain and subsist).

## Loss of Customary Rights to Land and Forest Products

As Colchester (2000) notes, plantations are often established in areas where local people have customary rights of access and resource use. They may also be established on lands owned by indigenous peoples, without their consent. This can lead to their loss of rights to land and resources, undermining traditional livelihoods and at its most extreme displacing people from rural lands to shantytowns or urban areas. Involved are the loss of physical access (via roads or trails), loss of legal access (people are forbidden to trespass on plantations or to use resources there), and loss of ecological access (ecological change brought about by the transformation of native habitat or farmland to plantation monocultures alters resource conditions and availability).

For example, in Madagascar communities have lost significant economic benefits when lands they were using for community-based, sustainable forest management were placed in large-scale logging concessions (whether sustainable or not), even though a large portion of the forest's value is in subsistence uses of nontimber forest products for artisanal production, food, fuel, fiber, and construction, all important to the quality of life of rural residents (Kremen et al. 2000).

Another example comes from India. Ravidranath et al. (1996) note that in India, where eucalyptus plantations are widespread, there is a loss of plant diversity on common and private lands. The loss deprives local communities of a flow of plant products traditionally obtained from natural forests and used for food, fodder, artisanal craft production, and oilseeds. Hiremath and Dandavatimath (1996) have documented this problem for Karnataka, illustrating a process that has occurred widely in India. There, the government leased large areas of forest and village common lands to pulp and paper corporations for establishing IMPIRs. Yet only a small percentage of rural families can subsist without using noncultivated common lands to supply fuelwood and animal fodder; and the landless are completely dependent on the commons to meet their biomass needs. The planting of eucalyptus and other fast-growing exotics on these common lands has deprived local communities of that source of fodder and fuel, undermining their subsistence. This alienation of village common lands has generated great conflict, as industry now competes with local communities for land. Local economies and cultures are being undermined, and seasonal and long-term migrations have become common.

A different response to the takeover of common lands by plantations has occurred in northern Morocco (Khattabi 1999). There, the population was originally nomadic shepherds. As eucalyptus plantations were established on common lands, the productive potential of pasture was reduced, reducing the availability of forage for livestock. This caused shepherds to settle in communities growing up around the plantations and take up employment in new professions associated with the production and processing of eucalyptus. Although Khattabi (1999) views this transition as a positive one and as an improvement socially, economically, and environmentally, that is not necessarily the case, nor is it clear that the local residents participated in the analysis or decisions.

## Displacement of Rural Residents

Some partisans in the intensive plantation/reserve debate typically state that IMPIRs can be established on marginal or abandoned croplands that have suffered ecological deterioration, helping to restore them. However, it should not be assumed that such marginal lands are uninhabited. Rather, marginal lands are often the only lands that the poor, politically powerless, and socially marginalized have access to. The displacement of indigenous peoples to marginal lands by European settlers and state governments has been widely documented throughout Africa and Latin America. Such "regions of refuge" (Clapp 1998) are areas where peasant agriculture may persist, more for subsistence than commercial purposes because of the marginal ecological quality of the land base. In such areas, plantation forestry may well out-compete peasant agriculture economically, so that displacement of rural populations is a major threat.

Clapp documents this process in south-central Chile. There, the indigenous Mapuche people are being displaced as corporations buy up state and private land and establish pine plantations to supply pulp mills. Landowners with small- and medium-sized holdings come under intense pressure to sell. As plantations are established, resident workers and tenant farmers are expelled and the land is fenced, with trespassing forbidden. Informal common property rights are eliminated, and people's relationships to land dissolve. The result is urban migration. A more gradual process of displacement occurs when peasants stay on their land, cultivating for subsistence but engaging in labor migration to plantations to earn income. As crop yields decline and wage labor increases, subsistence becomes more tenuous, and peasants become more likely to sell land to plantation corporations and migrate to urban areas (Clapp 1998). Similar processes threaten to unfold in neighboring Argentina, where the government is actively promoting the spread of plantation forestry (Aagesen 2001). When local people are viewed as competitors for land, as in these cases, they are likely to be displaced. The result is decay of indigenous culture and an increase in the population of urban poor (Clapp 1998).

IMPIRs are not only established in ecologically marginal areas and areas of abandoned cropland. It can be more economical to establish them in places that are easily accessible and have good-quality soil (Spencer and Jellinek 1995). For example, in Victoria, Australia, the government targeted good farmland for establishing eucalyptus plantations, trying to buy land from traditional farmers. The program caused heated opposition because it was perceived as accelerating an already existing decline in rural communities in Australia. As farmers sell land to government for plantations, the rural population declines, leading to declines in community services, neighborhood networks, and the viability of the remaining farms. Small farming communities are particularly vulnerable to the impact of plantations (Spencer and Jellinek 1995).

Oliver (1999) adds that not only do IMPIRs tend to increase urbanization, but so does the establishment of reserves. In both cases, people are displaced from lands they have traditionally occupied or used. Yet neither plantations nor reserves bring sufficient jobs and infrastructure development to keep people in the countryside.

Apart from the direct loss of rights to land and resources, the environmental impacts of plantations also affect local residents. These impacts can include the transformation of hydrological cycles, causing a loss of drinking water; the pollution of local water supplies by effluents from plantations and processing works; and in the tropics, changes in disease ecology, leading to the rise of malaria, dengue, scrub typhus, leishmaniasis, and filariasis (Colchester 2000).

In some cases, neighbors of forests may work out a social contract with the local forest industry, so that they receive some tangible benefits from that industry's forest management practices to help offset the negative impacts. However, it is unclear whether forest neighbors will find IMPIRS acceptable if the benefits all accrue outside the local area. For example, to what extent is a neighbor of a forest plantation likely to support forestry practices that are a potential threat to their health, or to local flora or fauna, based on the rationale that through some unidentified mechanism, "more important" biodiversity elsewhere will be protected? The widely held attitude summarized as "Not in my backyard" suggests that even when a community sees a particular activity as a public good it may be difficult to obtain support for it from neighbors. Arguments that a community should accept undesirable practices to free up land somewhere else for the benefit of other communities, or that other areas of the world have been identified by the scientific community as more deserving of biodiversity protection, may not be compelling to individuals in the local area where a plantation is to be established. If local communities are asked to bear risks, then the benefits to them should be clear.

## Effects of Protected Areas on Local People

Although we have argued here that IMPIRS are unlikely to generate reserves and protect native forests, we also acknowledge that within a specific area, a quid pro quo arrangement could be designed to promote that outcome. For example, a company might receive permission to establish IMPIRS on 200,000 hectares, as long as it maintained another 100,000 hectares in reserves. Another approach is for governments to assign forests within their countries to different zones—some to be used for IMPIRs, some as multiple-use areas, and some to be set aside as protected reserves (Binkley 1997). In the case of the latter, however, the question must be asked, What are these reserves to be protected from, and for whose benefit?

The idea of establishing forest reserves tends to imply that the forests in reserves will be protected from logging or deforestation for alternative land uses. In some cases, protected areas have successfully protected the environment and also supported traditional uses and sustainable livelihoods. But there is a continuum of management approaches for protected areas ranging from allowing local people to use forest products, such as plants and fuelwood, for subsistence, to prohibiting all human uses and entry at the most extreme. And even the most "protected" protected areas are not protected from human impacts such as air pollution or nonnative invasive species. As with plantations, the impact of reserves on local people will depend on what practices are acceptable in the protected area, including management activities (e.g., use of prescribed fire to restore certain plant species if they are endangered or of herbicides to remove exotic invasives).

There have been serious critiques of some approaches to the establishment of protected areas. Colchester (2000) observes that the dominant model has been to establish a network of strictly protected reserves, controlled by the state, cleared of resident communities, and under close government administration. But like plantations, protected areas are often located in places that local communities are inhabiting or using. Some of the local impacts of this model thus include curtailment or denial of resource access and use rights, loss of land rights, the undermining of local people's livelihood security, forced relocation, and the breakdown of indigenous knowledge and traditional resource use and management systems, all violations of human rights. The results are impoverishment, political marginalization, and the undermining of local social and cultural systems (Colchester 1994; CSQ 1985; Pimbert and Pretty 1995; Wells and Brandon 1992).

Serious questions have also been raised about the effectiveness of the exclusive reserve approach. When the land and livelihoods of local people are threatened by protected areas, they may respond by encroachment, poaching, sabotage, and unnecessary destruction of the natural resources and biodiversity the protected areas were designed to secure (Neumann 1998; Pimbert and Pretty 1995). New approaches are increasingly being used that recognize local peoples' rights and attempt to integrate the needs of local communities with the need to protect biodiversity (e.g., Western and Wright 1994). The Convention on Biological Diversity encourages state parties to respect, preserve, maintain, and protect indigenous peoples' traditional knowledge and cultural practices that are compatible with conservation or sustainable use, and some groups are working in a positive direction. Nevertheless, countries have been slow to put into practice the World Commission on Protected Areas 1999 guidelines, which emphasize comanagement of protected areas, agreements between indigenous people and conservation bodies, indigenous participation, and recognition of indigenous peoples' rights to "sustainable, traditional use" of their lands and territories.

Thus, it should not be assumed that establishing forest reserves will protect native forests. For the past six years, the International Union for the Conservation of Nature, World Commission on Protected Areas has been working on the issue of management effectiveness in protected areas. This was identified as a major global concern at the Fourth World Parks Congress in Caracas in 1992. Identified at the Fifth World Parks Congress, in Durban, South Africa, in 2001, was the concern that hunting and commercial trade in wildlife from protected areas across the tropics and subtropics are rapidly increasing and unsustainable, and many aspects are illegal (IUCN 2003).

Another threat to protected areas is illegal logging, and preventing it is of much interest to conservationists. For example, it has been estimated that virtually all logging of natural forests for export in India, Laos, Cambodia, Thailand, and the Philippines is illegal; that at least a third of the logging in Malaysia is illegal; and that up to 95% of the logging in Indonesia is not entirely legal (Poffenberger 1999). A realistic view of the world's forests must take into account the size of the gap between positive intentions and the reality as observed in practice, as policies are developed and implemented.

## *The Role of Local People in Decisions about Plantations and Reserves*

One reason for the ineffectiveness of protected areas in truly protecting nature has been the exclusion of local residents from decisionmaking regarding their management. Similarly, the failure to include local communities in decisionmaking relating to IMPIRs has led to heated opposition to them (Spencer and Jellinek 1995), as well as neglect and theft, reducing yields (Tapp 1996).

Central governments and private corporations typically maintain exclusive control over decisions regarding plantations and their management. Until recently, central governments have also maintained exclusive control over the management of protected areas. International nongovernmental organizations (NGOs) and scientists such as conservation biologists and ecologists have also participated in determining what should be protected in protected areas. In either case, control over management rests elsewhere than with local people. What should be the role of local communities in decisions about the management of forests, whether IMPIRs or reserves? Can including them in these decisions be beneficial?

Many people now recognize that without addressing issues of social and economic sustainability in local communities, ecological sustainability of native forests and protected areas is unlikely to be achieved. For example, the Fifth World Parks Congress developed recommendations on "good governance," which addressed the needs for recognition of diverse knowledge systems; openness, transparency, and accountability in decisionmaking; inclusive leadership; mobilizing support from diverse interests, with special emphasis on partners and local and indigenous communities; and sharing authority and resources and devolving/decentralizing decisionmaking authority and resources where appropriate (IUCN 2003).

Similar changes in attitudes have yet to occur in the arena of plantation establishment. Yet if IMPIRs are to be socially acceptable, and if they are to become increasingly widespread as one strategy for increasing the production of industrial roundwood, involving local communities in their establishment and management is critical to avoid negative social impacts such as we have described above. Local communities feel strongly about being involved in such decisions (e.g., Spencer and Jellinek 1995, Tapp 1996).

One approach would be to encourage local smallholders who live around IMPIRs to plant the same trees being grown on plantations, adopting them into their household production strategies (Clapp 1998). Diversification through integrating forestry and agriculture could help rural residents stay on the land and buttress economic stability. Another approach would be to address local people's needs and concerns when establishing IMPIRs.

Colchester (2000) cites Shell and World Wildlife Fund guidelines for eucalyptus plantations on peasant lands in northeastern Thailand as one example of this approach. The guidelines recommend the following measures:

- participatory rural appraisal methods to assess local realities,
- broad consultations with local people,
- careful impact assessments as part of the full planning cycle,
- evaluation of the implications for women,
- special attention to the resolution of potential conflicts over land tenure,
- open and continuing communications between the company and the communities,
- formal agreements with local communities incorporating guarantees on issues such as land allocation, forest protection, compensation for benefits foregone, employment and service contracts, marketing, infrastructural arrangements, and monitoring and evaluation, and

- mechanisms to share control of decisionmaking between local communities and the company.

The Shell/WWF guidelines also point out that commercial management strategies need to be developed for multispecies/multipurpose planting, which is more desirable for local people, building either on traditional forestry techniques or on existing agroforestry systems of intercropping, underplanting, or rotational cropping. They suggest experimenting with profit-sharing schemes and the establishment of landscape mosaics incorporating a mix of single species/purpose areas, multiple species/purpose areas, farmland, and other land uses (Shell/WWF 1993). Celiar (1998) cites participation, flexibility, empowerment, and commitment as the four characteristics needed when companies work with small farmers in contract farming for commercial woodlots of exotic species.

## Conclusions

Social contracts between communities of people and the management of the forests that surround them are negotiated through discussion, voluntary agreements, and international, national, and local policies. Unlike industrial roundwood, land and local inhabitants cannot be moved on world markets. Therefore, the local social contract is crucial to both communities and forests. The question of what international policies will foster both the protection and the use of forests is difficult to answer in a generalized way. It is open to debate among the many interest groups in the world forest policy community, the scientific community, the timber and pulp and paper industries, and the environmental community and organizations representing the rights of indigenous peoples.

In this chapter we have questioned many of the general statements and assumptions associated with the argument that intensive plantations can save native forests. We argue instead for framing the issue in terms of what forest practices are acceptable and appropriate on a given piece of land in a particular social context and who should decide that. We believe that plantations of various kinds and genetically engineered trees in those plantations, including those intensively managed for industrial roundwood production, will play an important role in the future. So will reserves designed to protect biodiversity. However, decisions for plantation and reserve establishment and management should take into account their advantages to local people, who should have a strong voice in developing the social contract for managing neighboring forests. Without that, IMPIRs may do little to contribute to environmental conservation.

# Notes

1. For a discussion of other types of plantations, especially agroforestry systems and community woodlots, see Long and Nair (1999).

2. We acknowledge that defining a "forest" is a difficult exercise and that forests represent a continuum of habitats having different densities of trees, different degrees of residential development, varying numbers and sizes of openings in the canopy cover, different uses, and so on. We adopt a broad definition of forests here, recognizing that there are specific definitions for different levels of inventory and assessment and ongoing discussions of preferred definitions for different purposes.

# References

Aagesen, D. 2001. Forest Policy in Chubut, Argentina: Lessons from the Upper Percey River Watershed. *Society and Natural Resources* 14(7): 599–607.

Adkoli, N.S. 1992. Social or Industrial Forestry: Industry's View. In *Growth and Water Use of Forest Plantations,* edited by I.R. Calder. New York: Wiley, 37–47.

Bandaratillake, H.M. 1996. Eucalyptus Plantations in Sri Lanka: Environmental, Social, Economic, and Policy Issues. In *Reports Submitted to the Regional Expert Consultation on Eucalyptus, Oct. 4–8, 1993. Volume 2,* edited by M. Kashio and K. White. Bangkok, Thailand: FAO Regional Office for Asia and the Pacific. RAP Publication 1996/44, 193–212.

Binkley, Clark S. 1997. Preserving Nature through Intensive Plantation Forestry: The Case for Forestland Allocation with Illustrations from British Columbia. *Forestry Chronicle* 73(5): 553–559.

Bowyer, J.L. 2001. Environmental Implications of Wood Production in Intensively Managed Plantations. *Wood and Fiber Science* 33(3): 318–333.

Brown, C. 2000. *The Global Outlook for Future Wood Supply from Forest Plantations.* Global Forest Product Outlook Study Working Paper No. GFPOS/WP/03. Rome, Italy: FAO, Forest Planning and Policy Division. http://www.fao.org/DOCREP/003/X8423E/X8423E00.HTM (accessed February 24, 2004).

Celiar, G.A. 1998. Small Scale Planted Forests in Zululand, South Africa: An Opportunity for Appropriate Development. *New Forests* 18(1): 45–58.

Clapp, R.A. 1998. Regions of Refuge and the Agrarian Question: Peasant Agriculture and Plantation Forestry in Chilean Araucania. *World Development* 26(4): 571–589.

Colchester, M. 1994. *Salvaging Nature: Indigenous Peoples, Protected Areas and Biodiversity Conservation.* UNRISD/World Rainforest Movement/WWF-International Discussion Paper No. 55. Geneva, Switzerland: United Nations Research Institute for Social Development.

————. 2000. *Indigenous People and the New "Global Vision" on Forests: Implications and Prospects.* Discussion paper for Global Vision 2050 for Forestry World Bank/WWF Project. http://greatrestoration.rockefeller.edu/21Jan2000/Colchester.htm (accessed February 24, 2004).

CSQ (Cultural Survival Quarterly). 1985. *Parks and People.* Volume 9:1.

Davis, T. 2000. *Sustaining the Forest, the People, and the Spirit.* Albany: State University of New York Press.

Doering, D. 2001. Personal communication with the authors.

Evans, J. 1992. *Plantation Forestry in the Tropics.* 2nd ed. Oxford, UK: Clarenden Press.

FAO (Food and Agriculture Organization of the United Nations). 2000. *Commodity Market Review 1999–2000.* Rome, Italy: FAO.

Haynes, R.W. 2003. *An Analysis of the Timber Situation in the United States: 1952 to 2050.* Gen. Tech. Rep. PNW-GTR-560. Portland, OR: USDA Forest Service, Pacific Northwest Research Station.

Hays, John P. 2001. Biodiversity Implications of Transgenic Plantations. In *Tree Biotechnology in the New Millennium: Proceedings of the International Symposium on Ecological and Societal Aspects of Transgenic Forest Plantations,* edited by S.H. Strauss and H.D. Bradshaw. Corvallis, OR: Oregon State University, 168–175. http://www.fsl.orst.edu/tgerc/iufro2001/symposia.htm (accessed February 24, 2004).

Helms, J.A. 1998. *The Dictionary of Forestry.* Bethesda, MD: Society of American Foresters.

Hiremath, S.R., and P.G. Dandavatimath. 1996. Eucalypt Plantations and Social and Economic Aspects in India. In *Reports Submitted to the Regional Expert Consultation on Eucalyptus, Oct. 4–8, 1993. Volume 2,* edited by M. Kashio and K. White. Bangkok, Thailand: FAO Regional Office for Asia and the Pacific. RAP Publication 1996/44, 58–66.

IUCN (International Union for the Conservation of Nature). 2003. *Fifth World Parks Congress Emerging Issues.* http://www.iucn.org/themes/wcpa/wpc2003/pdfs/outputs/wpc/emergingissues.pdf (accessed February 25, 2004).

Khattabi, A. 1999. Socio-economic Importance of Eucalyptus Plantations in Morocco. In *Global Concerns for Forest Resource Utilization: Sustainable Use and Management: Selected Papers from the International Symposium of the FORESEA MIYAZAKI 1998,* edited by A. Yoshimoto and K. Yukutake. Dordrecht, Netherlands: Kluwer, 73–82.

Kohlin, G., and P.J. Parks. 2001. Spatial Variability and Disincentives to Harvest: Deforestation and Fuelwood Collection in South Asia. *Land Economics* 77(2): 206–218.

Kremen, C., J.O. Niles, M.G. Dalton, G.C. Daily, P.R. Ehrlich, J.P. Fay, D. Grewal, and R.P. Guillery. 2000. Economic Incentives for Rain Forest Conservation across Scales. *Science* 288: 1828–1832.

Lautenschlager, R.A. 2000. Can Intensive Silviculture Contribute to Sustainable Forest Management in Northern Ecosystems? *Forestry Chronicle* 76(2): 283–293.

Long, A.J., and P.K. Nair. 1999. Trees outside Forests: Agro-, Community-, and Urban Forestry. *New Forests* 17(1-3): 145–174.

Maclaren, P. 2001. Forestry in New Zealand: The Opposite of Multiple Use? In *World Forests, Markets and Policies,* edited by M. Palo. Dordrecht, Netherlands: Kluwer, 365–370.

Mather, A.S. 1990. *Global Forest Resources.* London: Belhaven Press.

Neumann, Roderick P. 1998. *Imposing Wilderness: Struggles over Livelihood and Nature Preservation in Africa.* Berkeley: University of California Press.

Oliver, C.D. 1999. The Future of the Forest Management Industry: Highly Mechanized Plantations and Reserves or a Knowledge-Intensive Integrated Approach? *Forestry Chronicle* 75(2): 229–245.

Pandey, D. 1995. *Forest Resources Assessment 1990: Tropical Forest Plantation Resources.* Forestry Paper No. 128. Rome, Italy: FAO.

Pimbert, M.P., and J.N. Pretty. 1995. *Parks, People and Professionals: Putting "Participation" into Protected Area Management.* UNRISD Discussion Paper 57. Geneva, Switzerland: United Nations Research Institute for Social Development.

Poffenberger, Mark. 1999. *Communities and Forest Management in Southeast Asia: A Regional Profile of the WG-CIFM, Working Group on Community Involvement in Forest Management.* Gland, Switzerland: International Union for the Conservation of Nature.

Poffenberger, M., and B. McGean. 1996. *Village Voices, Forest Choices: Joint Forest Management in India.* Delhi: Oxford University Press.

Ravindranath, N.H., M. Gadgil, and J. Campbell. 1996. Ecological Stabilization and Community Needs: Managing India's Forest by Objective. In *Village Voices, Forest Choices: Joint Forest Management in India,* edited by M. Poffenberger and B. McGean. Delhi: Oxford University Press, 287–323.

Salwasser, Hal. 2001. Future Forests: Environmental and Social Contexts for Forest Biotechnologies. In *Tree Biotechnology in the New Millennium: Proceedings of the International Symposium on Ecological and Societal Aspects of Transgenic Forest Plantations,* edited by S.H. Strauss and H.D. Bradshaw. Corvallis, OR: Oregon State University, 10–19. http://www.fsl.orst.edu/tgerc/iufro2001/symposia.htm (accessed January 19, 2003).

Sargent, C., and S. Bass. 1992. *Plantation Politics: Forest Plantations in Development.* London: Earthscan.

Sedjo, R.A. 2001. The Economic Contribution of Biotechnology and Forest Plantations in Global Wood Supply and Forest Conservation. In *Tree Biotechnology in the New Millennium: Proceedings of the International Symposium on Ecological and Societal Aspects of Transgenic Forest Plantations,* edited by S.H. Strauss and H.D. Bradshaw. Corvallis, OR: Oregon State University, 29–46. http://www.fsl.orst.edu/tgerc/iufro2001/symposia.htm (accessed January 19, 2003).

Sedjo, R.A., and D.B. Botkin. 1997. Using Forest Plantations to Spare Natural Forests. *Environment* 30: 14–20.

Shell/WWF. 1993. *Guidelines: Shell/WWF Tree Plantation Review.* Shell International Petroleum Company Limited and World Wide Fund for Nature, London (cited from Colchester 2000).

Siry, J.P., F.W. Cubbage, and R.C. Abt. 2001. The Role of Plantations in World Forestry. In *Hardwoods—An Underdeveloped Resource? Proceedings of the Annual Meeting of the Southern Forest Economics Workers, Lexington, KY, March 26–28, 2000,* edited by M.H. Pelkki. Monticello: Arkansas Forest Resources Center, 105–111.

South, D.B. 1999. How Can We Feign Sustainability with an Increasing Population? In *Planted Forests: Contributions to the Quest for Sustainable Societies,* edited by J.R. Boyle. Forestry Sciences Volume 56. Dordrecht, Netherlands: Kluwer, 193–212.

Spencer, R.D., and L.O. Jellinek. 1995. Public Concerns about Pine Plantations in Victoria. *Australian Forestry* 58(3): 99–106.

Tapp, N. 1996. Social Aspects of China Fir Plantations in China. *Commonwealth Forestry Review* 75 (4): 302-308.

Victor, D.G., and J.H. Ausubel. 2000. Restoring the Forests. *Foreign Affairs* 79(6): 127–144.

———. 2001. Restoring the Forests. In *Tree Biotechnology in the New Millennium: Proceedings of the International Symposium on Ecological and Societal Aspects of Transgenic Forest Plantations*, edited by S.H. Strauss and H.D. Bradshaw. Corvallis, OR: Oregon State University, 47–56. http://www.fsl.orst.edu/tgerc/iufro2001/symposia.htm (accessed January 19, 2003).

Wear, D.N., and J.G. Greis. 2002. *Southern Resource Assessment—Summary Report.* General Technical Report SRS-53. Asheville, NC: USDA Forest Service, Southern Research Station.

Wells, M., and K. Brandon. 1992. *People and Parks: Linking Protected Areas with Local Communities.* Washington, DC: World Bank.

Western, David, and R.M. Wright. 1994. *Natural Connections: Perspectives in Community-Based Conservation.* Washington, DC: Island Press.

Whitmore, J.L. 1999. The Social and Environmental Importance of Forest Plantations with Emphasis on Latin America. *Journal of Tropical Forest Science* 11(1): 255–269.

# 10

# Have You Got a License for That Tree?

## (And Can You Afford to Use It?)

Nancy S. Bryson, Steven P. Quarles, and Richard J. Mannix

A number of natural resource and environmental laws potentially apply to the commercial production of transgenic trees in the United States. Some of them require premarket licensing by one or more federal agencies, including a comprehensive assessment of environmental impacts. Others provide the federal government, and sometimes citizens, with authority to take legal action to avert unreasonable risk to humans or the environment. Still others focus on special needs, such as those of endangered species or migratory birds.

In the post-StarLink era, greater attention is being paid to the way these laws relate to one another and to the range of legal requirements that can appropriately be used to address potential impacts on the environment and natural resources. Technology owners have also become painfully familiar with the liability inherent in license conditions that rely on user behavior. The licensing process needs to be transparent and protective without erecting a barrier to technology innovation.

## Federal Coordination

A consistent principle of health and environmental law in the United States is that products introduced into commerce should either be safe or, if not safe, present no unreasonable risk to people or the environment. How this principle is executed varies considerably depending on what law applies, which agency has jurisdiction, and the social perception of risk. For example, some products can be introduced only after government premarket review—drugs,

163

food additives, pesticides, and new chemicals. Sometimes express approval is a condition for product marketing; sometimes just "not saying no" is sufficient. Other products, such as consumer products containing hazardous substances, electrical equipment, appliances, and so forth, must meet performance standards but do not require preapproval. Generally, the intended use of the product determines which, if any, regulatory processes it must undergo or requirements it must satisfy.

Products of biotechnology do not always fit comfortably within the lines the law has drawn based on the historical function and intended use of products. A poplar tree that has been genetically modified to efficiently remove hazardous wastes from the environment may be subject to several types of regulatory review. The U.S. Department of Agriculture (USDA) has the responsibility of ensuring that the genetically modified tree is not a plant pest under the Plant Protection Act (PPA), which was enacted as part of the Agricultural Risk Protection Act of 2000. To the extent that a major federal action significantly affecting the environment is involved, application of the National Environmental Policy Act (NEPA) is triggered. NEPA compliance compels consideration of the impacts of, and alternatives to, the proposed action.

The Environmental Protection Agency (EPA) has promulgated regulations under the Toxic Substances Control Act (TSCA) that are applicable to "intergeneric microorganisms" that may be used to clean up wastes (TSCA, Sec. 2604). EPA defines "intergeneric microorganism" as "a microorganism that is formed by the deliberate combination of genetic material originally isolated from organisms of different taxonomic genera (*Code of Federal Regulations* [hereinafter *CFR*] 40, Sec. 725.3). Although EPA has not promulgated regulations for transgenic plants with similar capabilities, EPA's TSCA authority is also applicable to transgenic plants. Moreover, all aspects of a hazardous waste site remediation, including selection of the appropriate remedy, are closely regulated. For a product of biotechnology to get to market, each of the applicable legal standards must be met. The reviewing agencies look to science, data gaps, and default assumptions as the basis for risk evaluations and risk management decisions about commercialization.

The relationships among, and coordination of, these multiple authorities are governed by two policy statements developed by an interagency task group under the direction of the White House Office of Science and Technology Policy (OSTP 1986, 1992). Neither the 1986 Coordinated Framework for the Regulation of Biotechnology ("the framework") nor the 1992 Policy on Planned Introductions of Biotechnology Products into the Environment is a rule or creates any enforceable obligation. The documents simply describe how the different regulatory authorities are intended to fit together. The framework is "a first effort to aid in formulation of agency policy with respect to control of microorganisms developed by genetic engineering techniques"

(*Foundation on Economic Trends v. Johnson* 1986). The individual and collective experiences of the primary agencies of jurisdiction have also been translated into a substantial body of regulation and guidance for particular types of organisms. Each of the primary federal agencies has a Web site providing historical and current information on its activities in the regulation of biotechnology. These include: http://www.aphis.usda.gov/bbep (USDA); http://www.fda.gov/cber/reading. htm (FDA); and http://www.epa.gov/pesticides/biopesticides and http://www. epa.gov/opptinty/biotech/index.html (EPA).

# Applicable Law

In January 2001 the OSTP issued a series of case studies for public comment describing the path to market for six types of products of biotechnology based on the system of review that has matured under the framework. These case studies included genetically modified salmon, *Bt*-maize, herbicide-tolerant soybean, animals producing human drugs, bioremediation using poplar trees, and bioremediation and biosensing bacteria. They highlight the potential complexity of federal reviews of such products and the overlapping applicability of a number of the primary environmental and natural resource laws. The bioremediation poplar case study discusses the application and coordination of the Plant Protection Act, the Toxic Substances Control Act, and the Comprehensive Environmental Response, Compensation and Liability Act. (The case study is available at http://www.ostp.gov/html/012201.html.)

## *The Plant Protection Act*

USDA is authorized under the PPA to

> prohibit or restrict the importation, entry, exportation, or movement in interstate commerce of any plant, plant product, biological control organism, noxious weed, article, or means of conveyance, if the Secretary [of Agriculture] determines that the prohibition or restriction is necessary to prevent the introduction into the United States or the dissemination of a plant pest or noxious weed within the United States. (PPA, Sec. 7712(a))

The PPA consolidated USDA's authority under two previous acts, the Federal Plant Pest Act of 1957, and the Plant Quarantine Act of 1912. Effective June 22, 2000, those statutes were repealed and replaced by the PPA.

A "plant pest" is defined as any living stage of a protozoan, nonhuman animal, parasitic plant, bacterium, fungus, virus, infectious agent or pathogen, or similar or allied article that can directly or indirectly injure, cause damage to, or cause disease in any plant or plant product (PPA, Sec. 7702(14)). USDA views this concept broadly. The definition of "plant pest" covers "direct or indirect injury, disease, or damage not just to agricultural crops, but also to plants in general, for example, native species, as well as to organisms that may be beneficial to plants, for example, honeybees" (USDA 1997a).

The PPA is administered by USDA's Animal and Plant Health Inspection Service (APHIS), which has established regulations for the introduction of genetically modified plants, including trees, as "regulated articles." Regulated articles are defined generally as genetically engineered organisms and products that are derived from known plant pests. A regulated article is defined more specifically in the regulations as "any organism which has been altered or produced through genetic engineering, if the donor organism, recipient organism, or vector or vector agent belongs to any genera or taxa designated in subsection 340.2 [of the regulation]" and meets the definition of a plant pest, or any organism or product that APHIS determines or has reason to believe is a plant pest (*CFR* 7, Sec. 340.1). Organisms that are not classified may become regulated articles if there is reason to believe that they are plant pests. "Introduction" includes the importation, interstate movement, or "release into the environment" of a regulated article. Regulated articles are "released into the environment" when they are used outside the confinement of a laboratory, greenhouse, or other contained structure (*CFR* 7, Sec. 340.1).

The APHIS regulations allow for introduction by notification of all genetically modified plants that are not listed in the rules as noxious weeds (see, for example, *CFR* 7, Part 360) if the introduction meets certain eligibility criteria and performance standards. The eligibility criteria require, among other things, that the genetic material be "stably integrated" in the plant genome, that its function is known and that its expression does not result in plant disease, that it will not produce an infectious entity or be toxic to nontarget organisms, and that it has not been modified to contain certain genetic material from animal or human pathogens (*CFR* 7, Sec. 340.3(b)). The performance standards are essentially designed to ensure containment. Controls on shipment, storage, planting, identification, and the conduct and termination of the field trial are specified for this purpose (*CFR* 7, Sec. 340.3(c)).

If APHIS responds to the notification with a denial, or if the organism does not meet all of the regulatory provisions, a permit application may be submitted. Highlighted in the comments received during the APHIS rulemaking on these notification process provisions were certain concerns regarding the introduction of genetically modified trees. Cited as the basis for these concerns were lifespan in the field and the need to ensure reproductive confinement. APHIS acknowledged the existence of special issues for trees and included a requirement for annual renewal of notifications for all field tests exceeding one year from the date of introduction (USDA 1997b; *CFR* 7, Sec. 340.3(e)(4)).

## The National Environmental Policy Act

The National Environmental Policy Act (NEPA) requires that federal agencies publicly address the environmental impact of "every recommendation or report on proposals for legislation and other major Federal actions significantly affecting the quality of the human environment" (NEPA, Sec. 4332(2)(C)). Even for lesser federal actions involving more minor environmental effects, the federal agencies normally must "study, develop, and describe appropriate alternative recommended courses of action." (NEPA, Sec. 4332(2)(E)). NEPA compels agencies to develop methods and procedures that will ensure that unquantified environmental amenities and values be given appropriate consideration in decisionmaking. When the proposed agency action constitutes a "major Federal action" that has significant environmental consequences, the responsible federal official must prepare "a detailed statement" on the action, covering five specific issues, including environmental impacts of, alternatives to, and irreversible and irretrievable commitments of resources for the action (NEPA, Sec. 4332(2)(C)). This detailed statement is commonly referred to as an environmental impact statement (EIS).

Where analysis may be required to determine whether an agency action is a major federal action requiring an EIS, or to comply with the analytical requirements for a lesser federal action, the agency may initially prepare an environmental assessment (EA) that provides a brief discussion of the need for the proposed action, the alternatives, the environmental impacts, and the persons consulted. The EA will include sufficient analysis for determining whether to proceed with a more detailed EIS or to issue a "finding of no significant impact" (*CFR* 40, Sec.s 1501.4, 1508.9, 1508.13).

The EIS or EA serves as both a decisionmaking aid for the agency and an information repository for other interested parties. It provides environmental source material for evaluating the benefit of the proposed agency action in light of its environmental risks and for comparing these to the environmental risks presented by alternative courses of action. As the Supreme Court has stated:

NEPA has twin aims. First, it places upon an agency the obligation to consider every significant aspect of the environmental impact of a proposed action. Second, it ensures that the agency will inform the public that it has indeed considered environmental concerns in the decision making process. (*Baltimore Gas & Electric Co. v. Natural Resources Defense Council* 1983)

The sufficiency of analysis in EISs and EAs has been the subject of many lawsuits. NEPA does not require agencies to adopt any particular internal decisionmaking structure. Moreover, an agency is not constrained by NEPA from deciding that other values outweigh the environmental costs, as long as the environmental consequences have been fairly evaluated (*Vermont Yankee Nuclear Power Corp. v. Natural Resources Defense Council 1978*). Judicial decisions emphasizing the responsibility of the federal agency to take a hard look at the environmental impacts of its decisions are legion (*Kleppe v. Sierra Club 1976*). However, NEPA imposes no substantive obligation to take the least environmentally damaging alternative.

## The Toxic Substances Control Act

The Toxic Substances Control Act (TSCA) gives the EPA authority to regulate chemical substances that may present an unreasonable risk of injury to human health or the environment during their manufacture, processing, distribution in commerce, use, or disposal. A "chemical substance" is defined in the law to include "any organic or inorganic substance of a particular molecular identity, including

(i) any combination of such substances occurring in whole or in part as a result of a chemical reaction or occurring in nature and
(ii) any element or uncombined radical. (TSCA, Sec. 2602(2)(A)).

In a rule issued in April 1997, EPA interpreted this authority as extending to intergeneric microorganisms (*CFR* 40, Part 725). That regulation imposed a premarket review requirement on such microorganisms, with certain exemptions, and codified EPA's approach to research and development for microbial products of biotechnology. In proposing the rule, EPA noted that multicelled intergeneric plants and animals could also be chemical substances under TSCA. As a matter of policy, however, EPA limited this rulemaking to microorganisms and reserved its authority under TSCA to screen transgenic plants and animals in the future, as needed (U.S. EPA 1994). EPA's authority under TSCA includes

- premarket review of new chemical substances or significant new uses of existing chemical substances;
- regulation of existing chemical substances to avoid unreasonable risk; and
- recordkeeping/reporting requirements for health and environmental effects information (see, for example, TSCA, Sec. 2603, 2605, 2606, and 2607).

## The Comprehensive Environmental Response, Compensation and Liability Act

The Comprehensive Environmental Response, Compensation and Liability Act (CERCLA), also known as Superfund, established a multifaceted program to respond to the harm or potential harm to human health and the environment that hazardous substances may cause. EPA has developed a formal process to evaluate each release of a hazardous substance and each site where hazardous substances are located. The process includes a series of assessments that allow the primary threats to be identified, evaluated, and remedied. The most contaminated sites are placed on the National Priorities List, so as to focus limited resources on them. After carefully evaluating available remedial alternatives, EPA chooses the best one. Each CERCLA site is unique, and so is the remedial response.

The statutory goal of this process is to select the cleanup method that best protects human health and the environment, while complying with all applicable environmental laws, and that provides a long-term, cost-effective solution. During the selection process, EPA indirectly regulates bioremediation by identifying and evaluating the techniques that can be used to cleanse the site. The blueprint for all CERCLA response actions is set forth in the National Contingency Plan that is codified at Part 300 of *CFR* 40. The plan describes the procedures that EPA must follow when implementing the CERCLA program in situations where hazardous substances, pollutants, or contaminants have been released into the environment. Criteria for selection of the cleanup technique to be used at any particular site include community acceptance, as determined through a process of public notice and the solicitation and review of comments (*CFR* 40, Sec. 300.430(e)(9)(iii)).

The Resource Conservation and Recovery Act (RCRA) is a companion statute that is often used to require cleanup at sites where hazardous wastes are currently managed. The RCRA corrective action program functions in much the same way as CERCLA, with the difference, however, that RCRA requires the corrective action plan to be embodied in a permit issued by the appropriate federal or state agency.

## Case Study: The Bioremediation Poplar

As noted above, one of the case studies that the Office of Science and Technology Policy issued for comment examines the commercialization process for a hybrid poplar genetically engineered to detoxify trichloroethylene, or TCE. TCE is an industrial chemical that is chiefly used as a metal degreasing agent and as a solvent in paint strippers and lubricants. It is an environmental contaminant found throughout the industrialized world and a focus of remediation at many Superfund and RCRA corrective action sites.

This bioremediation poplar is described in the case study as still in the research and development stage, and not yet in field release trials, although APHIS notifications have been filed and accepted pursuant to the process described above. The federal agencies with primary jurisdiction are the USDA, using its authority under the PPA, and EPA, by virtue of its TSCA regulations. APHIS has expressed the opinion that permits will be required for the commercial use of these poplars; it is not planning to deregulate plants used in this application. In addition, the case study notes that trees grown for bioremediation purposes at Superfund and RCRA corrective action sites are expected to be subject to rigorous controls under those two statutes.

## The Tree and Its Function

The organism under the federal microscope in this case study is a hybrid between two cottonwood species: black cottonwood (*Populus trichocharpa*) and eastern cottonwood (*Populus deltoides*). Its key characteristics include:

- the ability to grow in soil with low levels of TCE,
- the ability to grow rapidly, and
- the ability to be vegetatively propagated.

The hybrid poplar was originally modified using a human cytochrome gene known for rapidly detoxifying TCE. A search is under way for other animal and plant genes that might perform the same function with greater public acceptability—a key issue in the success of ultimate commercialization. The tree is designed to absorb TCE from the soil or water in which it grows and express an enzyme that modifies the chemical into less toxic or nontoxic substances, which are translocated to the stems and leaves. The chemical transformation processes occur in the roots, stems, and leaves. There is no functional requirement for flowering. Application of techniques to ensure sterility limit the potential for outcrossing with wild relatives—an important safety consideration because the poplar is a species that outcrosses fairly easily.

The potential field of use of the bioremediation poplar is vast. There are thousands of TCE-contaminated sites where these trees could be employed for remedial purposes. On a typical site, stems would probably be grown and harvested on short rotations (five to seven years) sequentially, until the TCE had been reduced to target levels.

Depending on the metabolites in the stem tissue, the trees might be used in paper production. The hypothesis is that any organochlorine compounds remaining in stem tissue would be broken down by the chemical reactions in pulping. If such paper were to be produced, the Occupational Safety and Health Administration would regulate the safety of the manufacturing process. However, it is currently expected that steps will be taken at the test site to prevent gene flow and the persistence of the plants in the environment by destroying any vegetative parts that may remain, both above ground and below ground, after the test is completed.

## Adverse Effects of Potential Concern

The case study identifies four primary categories of risk associated with the bioremediation poplar:

- Health and ecological risks from the products of the TCE breakdown process on the remediation sites. (These include the potential for transpiration of TCE to air and the effects on insects and other organisms of metabolites.)
- The risk of escape of transgenes into the native eastern or black cottonwood populations or other reproductively compatible *Populus* species. (Both eastern and black cottonwoods are perennial, undomesticated plants expected to have wild relatives within pollination distance, and precautions to prevent escape of seeds or pollen are important.)
- Risk of transgenic trees themselves becoming weeds or otherwise invasive. (Could these trees effectively compete with their wild relatives in the environment?)
- Health and ecological risks from materials and products derived from the trees.

## Potential Benefits

TCE contamination is widespread, and TCE is a contaminant that is very difficult, expensive, time-consuming, and cumbersome to remove using existing techniques. Today, these techniques include pumping water from the aquifer and stripping the TCE by aeration or charcoal absorption. Other techniques use bacteria to degrade TCE but may require inducers such as toluene and phenol that present their own environmental and health risks.

Bioremediation poplars, on the other hand, could extract TCE while simultaneously contributing to the restoration of the ecological structure and function of the contaminated site. Existing poplars can withdraw TCE from their environment, but not very quickly. The purpose of the modification is to enhance the speed with which removal can be accomplished and produce a temporally effective form of remediation. Tree roots would prevent erosion. The trees themselves would provide humus and microclimatic changes, producing a more hospitable environment to many organisms. Poplars have a wide geographical distribution and could be successfully used in many areas. Propagation by cuttings is relatively easy and inexpensive. Poplars have an enormous water absorption capacity and could be used to clean shallow aquifers directly and deep aquifers indirectly through the irrigation use of pumped water.

## Legal Analysis

Although the case study does an excellent job of generally describing what the issues are for bioremediation poplars, as well as the review process employed to date, it is quite vague in its description of what comes next. Two phases of the current research and development process are described— proof of concept and successful bioremediation of contaminated soil. The first is under way. When a clone has been identified that detoxifies TCE and has acceptable growth characteristics, the proof-of-concept phase will end. Successful demonstration of bioremediation will require a large number of trees. The case study suggests that during the vegetative propagation stage appropriate coordination will occur among the various government agencies (identified as EPA, the Department of the Interior, the National Institutes of Health, the Forest Service in USDA, and others, as necessary), states, and tribal governments.

The case study includes a section on additional information and data to be generated, but it simply recites APHIS and EPA authority to require information. A section on mitigation, management, marketing, use, and disposal focuses on controlling reproduction and the need to completely destroy all vegetative parts at the end of the bioremediation project.

Reading between the lines, the case study suggests at least five legal issues in the commercialization of the bioremediation poplar that await resolution:

- the scope of EPA's role under TSCA;
- what conditions might reasonably be expected to be imposed through either an APHIS permit or a TSCA rule;
- whether a USDA permit for commercial release of the trees at a remediation site would trigger EIS requirements under NEPA;
- what if any Endangered Species Act issues might be presented by such a permit; and
- how such a release would fare under the CERCLA remedy selection criteria.

We evaluate each of these briefly in the balance of this chapter.

### EPA's Role under TSCA

The role of EPA under TSCA in this case study is unclear. What EPA has said to date is that it reserves its TSCA authority essentially as a legal gap-filling mechanism to address any unreasonable risk associated with the commercial release of genetically modified trees. Although EPA has issued explicit rules requiring premarket notification and review of intergeneric microorganisms, it has not yet adopted by rule an explicit premarket review requirement for multicelled organisms. The case for EPA jurisdiction under TSCA over transgenic plants and animals is also less than clear.

The potential authorities that EPA could exercise under TSCA in the absence of a premarket notification rule include a wide-ranging ability to:

- issue regulations to control unreasonable risks (including product bans),
- require manufacturers and processors to develop additional scientific data,
- secure imminent hazard injunctive relief in the federal courts, and
- impose record-keeping and reporting obligations.

TSCA also contains a citizen suit provision allowing "any person" to bring litigation to secure compliance with the provisions of the law and its implementing rules (TSCA, Sec. 2619(a)).

These authorities are quite general, and the courts have strictly construed EPA's rulemaking authority under TSCA to require the agency to choose the least burdensome and intrusive form of regulation (see, for example, *Corrosion Proof Fittings v. EPA* 1991). However, the case study clearly contemplates a role for EPA in tree commercialization and perhaps even in environmental release research experiments. There are references to EPA's activity with respect to microorganisms, described at some length in the companion case study on bioremediation and biosensing bacteria.

That case study discusses a field release experiment for biosensing bacteria conducted at the Oak Ridge National Laboratory in 1996 under a TSCA consent order. It describes a closely regulated experimental release and identifies a number of issues to be addressed prior to commercialization. In substance, these issues appear similar to those considered by APHIS, such as the stability of the new genetic construct. Issues of greater concern to EPA may include use of antibiotic resistance marker genes, the production of certain metabolites, and effects on nontarget species. Although the case study notes opportunities for coordination between the agencies, the mechanism for quick resolution of issues of competing jurisdiction is not apparent.

### Permit Conditions

The description of the benefits of using genetically modified trees to remediate hazardous waste sites is powerful. It is clear that, in concept, this use is highly regarded by the agencies and thought to be environmentally desirable. Equally evident is a significant discomfort with uncertainties about the degree of risk and the opportunities for risk control.

The case study stresses APHIS's intention always to require a permit for the commercial release of any trees genetically designed for remediation. It emphasizes that such trees will never be consumer products and that the potential for gene flow must be controlled by techniques designed to ensure that the bioremediation poplars do not flower. The commercialization of this tree will occur under the intense scrutiny generated by the national experience with StarLink corn. StarLink, a particular strain of *Bt* corn, was registered by EPA under the Federal Insecticide, Fungicide and Rodenticide Act (FIFRA), for animal feed only. It was not registered for human consumption because of regulatory concern about potential allergenicity to humans. The registration was conditioned, among other things, on grower agreements to ensure that StarLink corn would be marketed solely for animal feed.

These grower agreements proved to be ineffective in maintaining separation between the two commodity products. StarLink was found in several brands of taco shells and other grocery products containing yellow corn, prompting the product's recall. Numerous lawsuits were filed, the product's FIFRA registration has been voluntarily withdrawn, and the producer has since sold its agricultural chemicals division. In hindsight, reliance on grower agreements was a flawed concept and a commercial death sentence for the product.

The EPA employs registration conditions under FIFRA that rely on user behavior, including a requirement that growers plant appropriately sized refugia of nongenetically engineered crop in proximity to fields where the genetically modified crop is sown so as to reduce the potential for development of resistant insects and gene drift. The effectiveness of these conditions depends primarily on user compliance and self-policing, given the limited enforcement resources of the agencies. Many believe that even with full compliance with these conditions, resistant insect pests will develop. That is a political sore spot for organic and other farmers who rely on traditional *Bt* pesticides to a substantial degree.

It should be anticipated, therefore, that conditions dependent on human compliance for effectiveness will not be favored by either regulators or manufacturer/producers. The bioremediation poplar may present a unique situation in which pruning and other techniques could provide acceptable containment insurance against out-crossing because of the level of regulation of remediation activities. On the other hand, 30 years is a long time to rely on impeccable human control. Other alternatives of greater certainty are likely to be preferred, particularly by the communities situated near the sites where the genetically modified trees are grown.

To realize value from commercialization of these trees, the producer should think creatively about the types of permit conditions that can work in these unique applications. The law is open to innovative suggestion and interpretation, so long as the general performance standard of no unreasonable risk can be met.

## NEPA Compliance

The same considerations that cause APHIS to take the position that a permit will always be required for genetically modified trees and EPA to assert jurisdiction over genetically modified microorganisms suggest that a significant NEPA compliance requirement will apply to the commercialization of these trees. Although NEPA is a procedural statute and does not establish substantive criteria on which the agencies are to judge which alternative action to select, the case law does require a hard look by the agency at the environmental consequences of its action (*Kleppe v. Sierra Club* 1976).

The biotechnology arena has been exposed to NEPA litigation. Case law favors full-scale NEPA review (i.e., EISs) on "new" applications. For example, the National Institutes of Health could not approve a field release experiment involving bacteria modified to enhance frost resistance without an EIS (*Foundation on Economic Trends v. Heckler* 1985). The Department of the Interior was required to evaluate a new "bioprospecting" agreement involving Yellowstone National Park under NEPA procedures (*Edmonds Institute v. Babbitt* 1999). These cases suggest that any APHIS decision to permit the commercial release of bioremediation poplars will be preceded by NEPA compliance, most likely an EIS.

If EPA were to assume the dominant risk assessment function under TSCA, an argument might be made that NEPA should not apply to EPA's activities under the "functional equivalency" doctrine. Courts have provided an exception to the requirement that federal agencies must comply with NEPA "where a federal agency is engaged primarily in an examination of environmental questions [as in the case of decisions made under TSCA], and where substantive and procedural standards ensure full and adequate consideration of environmental issues" (*Warren County v. North Carolina* 1981). Such a process may be deemed the functional equivalent of NEPA review. However, that doctrine has been applied to TSCA decisions involving significant public participation, which typically would not be the case in a premarket notification process.

## The Endangered Species Act

The case study does not expressly address the Endangered Species Act (ESA). Since we are aware of no litigation concerning the applicability of the ESA to the federal licensing or private commercial production of transgenic trees, we simply provide an overview of the most relevant ESA provisions.

Under the ESA, species are determined to be endangered or threatened, and their critical habitat is designated, by rulemaking of the U.S. Fish and Wildlife Service in the Department of the Interior (FWS) and National Marine Fisheries

Service in the Department of Commerce (NMFS; "the service" or "services"). Four consequences arise from such ESA-implementing regulations.

First, ESA makes it unlawful for any public or private "person" to "take" an endangered species of fish or wildlife (ESA, Sec. 1538(a)(1)(B)). The ESA also makes it unlawful to damage or destroy endangered plants on federal lands, or to do the same on private lands if done in knowing violation of state law or in violation of state criminal trespass law. Exercising the discretionary authority provided by the ESA, the services have generally extended the "take" prohibition to "threatened" fish and wildlife species (*CFR* 50, Sec. 17.31(a)). The term "take" means "to harass, harm, pursue, hunt, shoot, wound, kill, trap, capture, or collect, or to attempt to engage in any such conduct" (ESA, Sec. 1532(19)). The regulatory definitions of "harm" include even an innocent land use activity ("significant habitat modification or degradation"), if that activity "actually kills or injures [endangered or threatened] wildlife" (*CFR* 50, Sec. 17.3, 222.102; *Babbitt v. Sweet Home Chapter of Communities for a Great Oregon* 1995). Thus, to the extent that an endangered or threatened bird species would be killed or injured, directly or perhaps through failed reproduction, by, for example, making a nest of materials from the bioremediation poplars, "harm" may occur.

Second, the ESA imposes (a) the substantive limitations that each federal agency must "insure that any action authorized, funded, or carried out by such agency . . . is not likely to jeopardize the continued existence of any endangered species or threatened species or result in the destruction or adverse modification of" any critical habitat designated for that species; and (b) the requirement that the federal agency taking the action assess the jeopardy and critical habitat threats "in consultation with" the relevant service. (ESA, Sec. 1536(a)(2)). Federal agency actions concerning transgenic trees would be subject to these consultation requirements.

The services have adopted regulations governing their consultation role (*CFR* 50, Part 402). Basically, the rules: (a) do not require any consultation procedures if the federal agency determines that its action would have no effects on an endangered or threatened species or critical habitat; (b) allow an expedited informal consultation if the action agency finds, with the written concurrence of the relevant service, that the action is not likely to adversely affect the species or critical habitat; and (c) require formal consultation, with preparation of a biological evaluation by the action agency and a biological opinion by the relevant service, if the agency action is likely to adversely affect the species or critical habitat (*CFR* 50, Sec. 402.13-402.14; see ESA, Sec. 1536(b)).

Third, in the 1982 ESA amendments, Congress provided two mechanisms for authorizing some limited "take" of endangered or threatened wildlife incidental to an otherwise lawful land use or other activity. For actions that require some form of federal authorization, Congress prescribed that, where an action conforms to the standard against jeopardizing the continued existence of an endangered or threatened species or destruction or adverse modification of the critical habitat, the incidental "take" of a few individuals may be allowed by an "incidental take statement" issued with the service's biological opinion, if the federal agency and the permittee follow the "reasonable and prudent measures" that the service prescribes in the opinion (ESA, Sec. 1536(b)(4); CFR 50, Sec. 402.14(i)). For nonfederal actions that have no access to the consultation process, Congress allowed the services to grant incidental take permits if the applicant agrees to commit resources to a habitat conservation plan and if certain other standards are met (ESA, Sec. 1539(a)(2)). These provisions provide potential safe harbors against, or immunity from, violation of the "take" prohibition associated with the agency actions.

Fourth and finally, ESA provides three forms of enforcement or relief for ESA "take" or consultation violations. Continuing violations can be enjoined either in ESA citizen suits brought by private parties or in injunctive relief actions pursued by federal authorities. Federal authorities also can seek civil and criminal penalties for a "take" or other ESA violations (ESA, Sec. 1540(a) and (b)). For the bioremediation poplar, efforts should be made to avoid applicability of the ESA by conducting field trials and ultimately employing the tree at corrective action sites that are unlikely to serve as habitat for endangered or threatened species. This may be difficult to achieve, as the poplar itself may serve as habitat for endangered or threatened bird species. To avoid "take" liability for the producers and growers of the bioremediation poplar, APHIS should seek the protection of incidental take statements whenever formal consultations occur on permits for the tree. If no consultation is undertaken, growers may wish to prepare habitat conservation plans and apply for incidental take permits to secure "take" immunity.

## CERCLA Remedy Selection

EPA regulations set nine criteria for selecting among the alternative remedies for a Superfund site:

- overall protection of human health and the environment;
- compliance with applicable or relevant and appropriate requirements under federal and state environmental law;
- long-term effectiveness and permanence;
- reduction of toxicity, mobility, or volume through treatment;
- short-term effectiveness;
- implementability;
- cost;
- state acceptance;
- community acceptance. (CFR 40, Sec. 300.430(e)(9)(iii)).

The last criterion is likely to be the most significant hurdle to realizing the value of the commercialization of the bioremediation poplars. As the education effort will be substantial, it cannot begin too soon.

# Conclusion

A number of legal uncertainties lie on the road to market for bioremediation poplars. Other genetically modified trees will face different issues, depending on their intended use. Fruit trees will be subject to FDA jurisdiction to review the safety of the modified fruit as a food. Trees engineered to absorb extra carbon as they grow will likely need to be qualified for measurable and enforceable reductions in carbon dioxide if they are to be used for greenhouse gas control, consistent with requirements for ozone control reduction measures in the current Clean Air Act (CAA). Trees designed with different lignin structures to enhance the ease of paper production will almost certainly face ESA issues.

Scientific advances in plant genomics are expected to accelerate over the next decade. The likely response of the federal government will be to strengthen various aspects of its oversight of agricultural biotechnology. Indeed, in February 2002 a committee of the National Research Council (NRC) released a report that considered the potential hazards of introducing transgenic plants into the environment and recommended that USDA apply a more rigorous review of new transgenic plants before approving them for commercial use. The report, which was commissioned by USDA, concluded that the USDA regulatory process needs to be made "significantly more transparent and rigorous by enhanced scientific peer review, solicitation of public input, and development of determination documents with more explicit presentation of data, methods, analyses, and interpretations." Whenever changes in regulatory policy are being considered, USDA should convene a scientific advisory group (as EPA typically does) and before making specific, precedent-setting decisions should solicit broad, external scientific review, well beyond the use of *Federal Register* notices, to which few seem to respond.

In August 2002, OSTP, working with USDA, EPA, and FDA, reacted to the NRC report by requesting public comment on the outline of a series of coordinated federal actions to, among other things, update field testing requirements (OSTP 2002). As we wrote this chapter, specific proposed rules had not been published. Initially, the proposed measures will apply only to biotechnology-derived crops for food and feed. Actions addressing other regulatory aspects of biotechnology-derived plants will be proposed in the future, as efforts to enhance public confidence in the regulatory oversight of biotechnology proceed. Any strengthening of requirements within the existing, "coordinated" legal structure should avoid unduly complicating progress toward commercialization. If exhaustive procedures or aggressive standards are established, Congress may intervene. In any case, understanding the federal legal system and charting a path to the market through that system will be crucial to commercial success.

# References

*Babbitt v. Sweet Home Chapter of Communities for a Great Oregon,* 515 U.S. 687 (1995).

*Baltimore Gas & Electric Co. v. Natural Res. Def. Council,* 462 U.S. 87, 97 (1983).

CAA (Clean Air Act). *U.S. Code* Title 42. Chapter 85. Sections 7401–7671q. (2000).

CERCLA (The Comprehensive Environmental Response, Compensation and Liability Act). *U.S. Code* Title 42. Chapter 103. Sections 9601–9675 (2000).

*Corrosion Proof Fittings v. EPA,* 947 F.2d 1201 (5th Cir. 1991).

*Edmonds Institute v. Babbitt,* 42 F. Supp. 2d 1, 17-20 (D.D.C. 1999).

ESA (Endangered Species Act). *U.S. Code* Title 16. Chapter 35. Sections 1531-1544 (2000).

FIFRA (Federal Insecticide, Fungicide, and Rodenticide Act). *U.S. Code* Title 7. Sec. 136 et seq. (2000).

*Foundation on Economic Trends v. Heckler,* 756 F.2d 143 (D.C. Cir. 1985).

*Foundation on Economic Trends v. Johnson,* 661 F. Supp. 107, 109 (D.D.C. 1986).

*Kleppe v. Sierra Club,* 427 U.S. 390, 410 n.21 (1976).

NEPA (National Environmental Policy Act). *U.S. Code* Title 42. Sections 4321-4347 (2000).

(NRC) National Research Council. Committee on Environmental Impacts Associated with Commercialization of Transgenic Plants. 2002. *Environmental Effects of Transgenic Plants: The Scope and Adequacy of Federal Regulation.* Washington, DC: National Academy Press.

(OSTP) Office of Science and Technology Policy. 1986. Notice: Coordinated Framework for Regulation of Biotechnology. *Federal Register* 51(June 26): 23302.

———. 1992. Notice: Exercise of Federal Oversight within Scope of Statutory Authority: Planned Introductions of Biotechnology Products into the Environment. *Federal Register* 57(Feb. 2): 6753.

———. 2002. Notice: Proposed Federal Actions to Update Field Test Requirements for Biotechnology Derived Plants and to Establish Food Safety Assessments for New Proteins Produced by Such Plants. *Federal Register* 67(Aug. 2): 50578.

PPA (Plant Protection Act). *U.S. Code* Title 7. Sec. 7701–7772 (2000).

RCRA (Resource Conservation and Recovery Act). *U.S. Code* Title 42. Sec. 6901–6992k (2000).

TSCA (Toxic Substances Control Act). *U.S. Code* Title 15. Sec. 2601–2692 (2000).

USDA (U.S. Department of Agriculture, Animal and Plant Health Inspection Service). 1997a. Notice: Monsanto Co.; Receipt of Petition for Determination of Nonregulated Status for Genetically Engineered Tomato. *Federal Register* 62 (Nov. 28): 63312.

―――. 1997b. Final Rule: Genetically Engineered Organisms and Products; Simplification of Requirements and Procedures for Genetically Engineered Organisms. *Federal Register* (May 2): 23945, 23947.

U.S. EPA (Environmental Protection Agency). 1994. Proposed Rule: Microbial Products of Biotechnology; Proposed Regulation under the Toxic Substances Control Act. *Federal Register* 59 (Sept. 1): 45526, 45527.

*Vermont Yankee Nuclear Power Corp. v. Natural Resources Defense Council*, 435 U.S. 519, 540–48 (1978).

*Warren County v. North Carolina*, 528 F.Supp. 276, 286 (E.D.N.C. 1981).

# 11

# Invasiveness of Transgenic versus Exotic Plant Species
## *How Useful Is the Analogy?*

James F. Hancock and Karen Hokanson

It has commonly been suggested that invasive, exotic species can be used as models for evaluating the risk of release of transgenic crops (NAS 1987; Tiedje et al. 1989; Parker and Kareiva 1996; Marvier 2001). We will argue in this chapter that that analogy is not valid.

We are all familiar with the "environmental disasters" associated with the introduction of exotic species. In many instances, the problem species were intentionally introduced, for example, rhododendron in the United Kingdom, pine in Australia, kudzu in the southeastern United States, and purple loosestrife in eastern North America (Keeler 1988; Mooney and Drake 1986; Crawley 1997). Others were introduced accidentally, such as the Dutch elm disease and corn leaf blight in North America. The vast majority of introduced organisms perish or do not establish self-sustaining populations (Pimentel et al. 1989), but we keep being drawn to those that do.

## Characteristics of Invasive Species

What does make a species invasive? To answer this question, we first need to define what we mean by "invasive." Probably the most common definition given is, having the ability to increase when rare; however, all successful species meet this criterion. Crawley (1997) suggests that invasive species should really be called "problem plants" when the species has passed some threshold of abundance and someone is concerned. He suggests that to understand the population biology of an invading plant genotype, we need knowledge of the following: (a) abiotic environment, (b) the biotic environment, (c) interaction between biotic and abiotic environment, and (d) the year. In other words, invasiveness is very complex.

Sarah Reichard recently published a series of papers presenting a framework for evaluating plant invasiveness (Reichard and Cambell 1996; Reichard and Hamilton 1997; Reichard 1999). Her decision tree is based on a predictive model derived from discriminant and regression analysis of a number of structural, life history, and biogeographical characteristics of introduced woody plants. The characteristics analyzed in those studies, most of which have routinely been associated with invasiveness, included native range, whether or not the species invades elsewhere, leaf longevity, polyploidy, reproductive system, vegetative reproduction, minimum juvenile period, length of flowering period, flowering season, length of the fruiting period, fruiting season, dispersal mechanism, seed size, and seed germination requirements.

Of the woody plants that have invaded the United States, Reichard found that 54% invade other parts of the world, 44% spread by vegetative means, 51% have seeds that germinate without pretreatment, and 1% are interspecific hybrids. Only 3% have been introduced from other parts of North America. Based on these results, Reichard developed a decision tree for acceptance of exotic woody species into North America, which begins with the question, Does the species invade elsewhere, outside of North America? Two other important questions in the decision tree are, Is the species in a family or genus with species that are already strongly invasive in North America? and, Is the species native to parts of North America other than the region of the proposed introduction? Other questions concern whether or not the species is a sterile interspecific hybrid, its rate of vegetative reproduction, the length of the juvenile period, and germination requirements.

What Reichard's analysis indicates is that a high percentage of the exotic species that become invasive are already excellent colonizers somewhere else, and their population size explodes when they are introduced into a new area where there are few to none of the natural constraints with which they evolved. This is very different from the situation facing transgenic forestry and agronomic crops. They will not be removed from the complex array of natural constraints that currently face them, and only a very limited number of those constraints will be removed by the addition of a new trait through genetic engineering. The array of factors regulating natural populations must be complex, as the introduction of single biological control agents has rarely had much of an impact on invasive, exotic species (Pimentel et al. 1984).

## Invasiveness of Agronomic and Forestry Species

Only a small percentage of agronomic and forestry crops are important weeds outside of agro-environments (Table 11-1). They rely on human dis-

**TABLE 11-1.** Survival of North American Crops in Native Environments

| Nonpersistent | Persistent/noninvasive | Persistent/invasive |
|---|---|---|
| Beet | Apple | Barley |
| Broccoli | Asparagus | Rapeseed (Canola) |
| Carrot | Blueberry | Rice |
| Cauliflower | Cranberry | Sorghum |
| Celery | Pear | Sunflower |
| Citrus | Poplar | Wheat |
| Cucumber | Spruce | |
| Cotton | Strawberry | |
| Eggplant | | |
| Lettuce | | |
| Maize | | |
| Melon | | |
| Onion | | |
| Pea | | |
| Peanut | | |
| Pepper | | |
| Potato | | |
| Soybean | | |
| Squash | | |
| Sugarcane | | |
| Sunflower | | |
| Tobacco | | |
| Watermelon | | |

*Source:* Hancock, J.F., R. Grumet, and S.C. Hokanson, 1996. The Opportunity for Escape of Engineered Genes from Transgenic Crops. *HortScience* 31: 1080–1085.

turbances to become established and rarely persist outside of specific habitats. Clearly exceptions exist, such as barley, rapeseed, and rice, but over 80% of all crop species do not persist in native environments. Crawley et al. (2001) have generated some excellent evidence of how poorly crop genotypes do in native environments whether they are genetically modified or not. When they compared the performance of transgenic and nontransgenic rape, maize, beet, and potato in 12 native environments, the genetically modified plants were never found to be more invasive or persistent than their antecedents. In fact, all populations of maize, rape, and beet were extinct after four years, and only conventionally bred potatoes were left after 10 years (and only at one site). The transgenic rape and maize expressed tolerance to the herbicide glufosinate, the genetically modified sugar beet was resistant to glyphosate, and the transgenic potatoes expressed either the insecticidal *Bt* toxin or pea lectin gene.

In his classic work, Baker (1965, 1974) associated a complex array of traits with colonizing ability, including broad germination requirements, short and long seed dispersal, discontinuous germination, long-lived seed, vigorous vegetative reproduction, rapid growth to flowering, brittle propagules, continuous seed production, vigorous competitors, self-compatibility, unspecialized pollinators, very high seed output, plastic seed production, and polyploidy. When Keeler (1989) took Baker's weediness traits and compared the worst weeds with agronomic crops, she found that serious weeds possessed an average of 81% of these traits, random non-weeds had 59%, and crop plants had 42%.

To date, 11 tree crops have been genetically engineered in the United States and tested in the field: apple, papaya, citrus, persimmon, pear, plum, pine, poplar, sweetgum, spruce, and walnut. When rated according to Baker's characteristics, all fall well below the random non-weeds, displaying from 21% to 50% of the traits (Table 11-2). Poplar has the highest average (50%), possessing the weediness traits unspecialized pollinators, variable seed dispersal distance, high seed production, seed production in many environments, vigorous vegetative propagation, brittle propagules, and polyploidy. However, they are outcrossing; have discontinuous seed production, short seed longevity, narrow germination requirements, and discontinuous germination; are weak competitors; grow slowly; and are only rarely polyploid.

This suggests that in most agronomic and forestry crops, a whole syndrome of traits would need to be altered through genetic engineering to make them invasive. Because agronomic crops are often poor competitors in nature, their impact on native populations has also been generally limited to introgression. There are numerous instances where hybridization with wild relatives has increased the weediness of the native species in agronomic fields through crop mimicry (Ellstrand et al. 1999), but there is little evidence of crop genes affecting the overall fitness of a native species. Even though crop species have been planted among their progenitors for thousands of years, we are not aware of any report where the native fitness of the wild species was noticeably changed. When David Duvick (2000) asked 20 experienced plant breeders if the introduction of conventional resistance genes has led to undesirable consequences with respect to the weediness of a crop or its relatives, the breeders knew of no example.

## Predicting the Environmental Risk of Genetically Modified Organisms

It has been suggested that genetically engineered trees pose significantly greater environmental risks than do genetically engineered food crops

TABLE 11-2. Weediness Traits in Transgenic Trees That Have Been Field Tested in the United States

| Weediness trait | Apple | Papaya | Citrus | Persimmon | Pear | Plum | Pine | Poplar | Sweetgum | Spruce | Walnut |
|---|---|---|---|---|---|---|---|---|---|---|---|
| Broad germination requirements | no | no | no | no | no | no | no | no | no | no | no |
| Discontinuous germination | no | no? | no | no | no | no | no | no | no | no | no |
| Long-lived seeds (>five years) | no | no | no | yes | no | no | yes | no | yes | yes | yes |
| Rapid growth | no | yes | no | no | no | no | no | no | no | no | no |
| Continuous seed production | no | no | no | no | no | no | no | no | no | no | no |
| Self pollinated | no | no | yes | no | no | no | no | no | no | no | no |
| Unspecialized pollinators | no | yes | no | no | no | no | yes | yes | yes | yes | no |
| High seed output | yes | yes | yes | yes | yes | yes | yes | yes | yes | yes | yes |
| Seeds produced in many habitats | yes | no | no | yes | yes | yes | yes | yes | no | yes | no |
| Short and distant seed dispersal | yes | yes | yes | yes | no | yes | yes | yes | yes | no | no |
| Vigorous vegetative reproduction | no | no | no | no | no | no | no | yes | yes | no | no |
| Brittle propagules | no | no | no | no | no | no | no | yes | no | no | no |
| Vigorous competitors | no | no | no | no | no | no | no | no | no | no | no |
| Polyploid ($2n > 28$) | yes | no | no | yes | yes | no | no | yes | yes | no | yes |
| Percentage of traits | 28 | 28 | 21 | 36 | 28 | 21 | 36 | 50 | 43 | 28 | 21 |

because the genes inserted into trees are more likely to "escape" into the wider environment (Campbell 2000). Plantation trees have been altered through breeding far less than most agronomic crops and as a result are much more closely adapted to native habitats than most crop species. However, most are not highly invasive in their native geographic range, and the transgenic derivatives and any native/engineered hybrids will be subjected to the complex array of factors that normally regulate the native populations. The bottom line in assessing the environmental risk of both transgenic trees and transgenic herbaceous crops is the nature of the transgene, that is, how significant an impact it will have on the fitness of native populations should it escape.

In fact, it is much easier to predict the environmental risk of transgenic trees than that of an exotic introduction, as the level of risk in transgenics can be measured by evaluating the fitness impact of a single engineered trait, rather than a whole syndrome of potentially invasive traits. A unique genotype is not being introduced into an environment where its native constraints are removed. The species is already in the environment, and we know how invasive it is. What we need to worry about is whether the addition of a single gene will increase its existing level of invasiveness to problem levels. An increase in vegetative reproduction, were this to occur without sacrificing sexual reproduction; a decrease in the pretreatment requirements of seeds, were this not to reduce seedling survival; or a shortened juvenile period, were this not to reduce ability to grow tall and compete for light, could certainly raise red flags concerning invasive potential. But these alterations are difficult to achieve and currently no more likely to be accomplished through genetic engineering than they are through traditional genetic improvements.

In some cases, the risk involved in the deployment of transgenes can be efficiently evaluated through the concept of familiarity (Hokanson et al. 2000). The Animal and Plant Health Inspection Service now assesses risk based on the biology of the crop, the nature of the introduced trait, the receiving environment, and the interaction among them. Knowledge of these factors provides familiarity, which allows decisionmakers to compare genetically engineered plants to their nonengineered counterparts. Familiarity allows regulators to efficiently assign levels of risk, without doing any additional experiments, when the phenotypic effects of transgenes closely mimic conventionally deployed or native genes. Hokanson et al. (2000) have outlined a number of examples in which transgenic genotypes have similar nontransgenic phenotypes, such as insect and virus resistance.

This approach was the one recommended by the first group of scientists who evaluated the environmental risks of transgenic crops. In their often cited paper, Tiedje et al. (1989) state, "[T]ransgenic organisms should be evaluated and regulated according to their biological properties (phenotypes), rather than according to the genetic techniques used to produce them...." They add, "Long term experience derived from traditional plant breeding provides useful information for the evaluation of genetic alterations similar to those that might have been produced by traditional means, and such alterations are likely to pose few ecological problems." One of the major conclusions of the National Academy of Sciences report *Field Testing Genetically Modified Organisms: Framework for Decisions* (1989) was that crops modified by genetic engineering will pose risks that are no different from those modified by classical genetic methods.

The problem with using the concept of familiarity is finding genes of equivalent effect and strength in natural populations. Reasonable arguments can be made for many of the transgenes that are similar to conventionally deployed resistance genes, but numerous other engineered genes will produce phenotypes that are unique to the species or have broader effects than the native genes. Some of these transgenes, such as herbicide resistance, are likely to be effectively neutral in the native environment, but others that alter reproductive potential and physiological tolerances may have much more significant impacts. Regardless, it is much easier to assign risk to transgenic crops than to exotic species, as we can restrict our worry to the effect of one gene on the fitness of a species in the place it is already grown, rather than making guesses about the fitness of a whole species genome in a unique environment.

## Conclusions

The patterns by which invasive, exotic plant species have spread cannot be used to predict the environmental impact of transgenic trees and agronomic crops. Although it is true that some transgenes will influence individual traits associated with invasiveness, numerous other natural characteristics of these species make single changes unlikely to substantially alter their competitiveness. Invasive species have almost always been introduced in places where they have few to none of the natural constraints with which they evolved. The species then fills a new niche and its population numbers explode. In many cases, the species were already invasive in their original habitat. This is very different than making a single change in a species already subject to multiple natural controls. Most engineered species are poor colonizers and will be grown in their original environment, with its complex array of natural constraints. Normally, only one of these constraints will be removed by the addition of a new trait through genetic engineering. In contrast to exotic species, the risk of deploying transgenes can be effectively predicted by considering the phenotype conferred by a single gene and the overall invasiveness of the crop itself.

# References

Baker, H.G. 1965. Characteristics and Modes of Origin of Weeds. In *The Genetics of Colonizing Species*, edited by H.G. Baker and G.L. Stebbins. New York: Academic Press.

————. 1974. The Evolution of Weeds. *Annual Review of Ecology and Systematics* 5: 1–23.

Campbell, F.T. 2000. *Genetically Engineered Trees: Questions without Answers.* Washington, DC: American Lands Alliance.

Crawley, M.J. 1997. Biodiversity. Chapter 19 in *Plant Ecology*, edited by M.J. Crawley. Oxford, UK: Blackwell Science Ltd.

Crawley, M.J., S.L. Brown, R.S. Hails, D.D. Kohn, and M. Rees. 2001. Transgenic Crops in Natural Habitats. *Nature* 409: 682–683.

Duvick, D.N. 2000. Consequences of Classical Breeding for Pest Resistance. In *Ecological Effects of Pest Resistance Genes in Managed Ecosystems*, edited by P.L. Traynor and J.H. Westwood. Blacksburg, VA: Information Systems for Biotechnology, 37–42.

Ellstrand, N.C., H.C. Prentice, and J.F. Hancock. 1999. Gene Flow and Introgression from Domesticated Plants into Their Wild Relatives. *Annual Review of Ecology and Systematics* 30: 539–563.

Hancock, J.F., R. Grumet, and S.C. Hokanson, 1996. The Opportunity for Escape of Engineered Genes from Transgenic Crops. *HortScience* 31: 1080–1085.

Hokanson, K., D. Heron, S. Gupta, S. Koehler, C. Roseland, S. Shantharam, J. Turner, J. White, M. Schechtman, S. McCammon, and R. Bech. 2000. The Concept of Familiarity and Pest Resistant Plants. In *Ecological Effects of Pest Resistance Genes in Managed Ecosystems*, edited by P.L. Traynor and J.H. Westwood. Blacksburg, VA: Information Systems for Biotechnology, 15–20.

Keeler, K.H. 1988. Can We Guarantee the Safety of Genetically Engineered Organisms in the Environment? *CRC Critical Reviews in Biotechnology* 8: 85–97.

————. 1989. Can Genetically Engineered Crops Become Weeds? *Bio/technology* 7: 1134–1139.

Marvier, M. 2001. Ecology of Transgenic Crops. *American Scientist* 89: 160–167.

Mooney, H.A., and J.A. Drake. 1986. *Ecology of Biological Invasions of North America and Hawaii.* New York: Springer-Verlag.

NAS (National Academy of Science). 1987. *Introduction of Recombinant DNA-Engineered Organisms into the Environment: Key Issues.* Washington, DC: National Academy Press.

Parker, P., and I.M. Kareiva. 1996. Assessing the Risks of Invasion for Genetically Engineered Plants: Acceptable Evidence and Reasonable Doubt. *Biological Conservation* 78: 193–203.

Pimentel, D.C., C. Glenister, S. Fast, and D. Gallahan. 1984. Environmental Risks of Biological Pest Control. *Oikos* 42: 283–290.

Pimentel, D., M.S. Hunter, J.A. LaGro, R.A. Efroymson, J.C. Landers, F.T. Mervis, C.A. McCarthy, and A.E. Boyd. 1989. Benefits and Risks of Genetic Engineering in Agriculture. *BioScience* 39: 606–614.

Reichard, S.H. 1999. A Method for Evaluating Plant Invasiveness. *Public Garden* 14: 18–21.

Reichard, S., and F. Cambell. 1996. Invited but Unwanted. *American Nurseryman* 184: 39–45.

Reichard, S.H., and C.W. Hamilton. 1997. Predicting Invasions of Woody Plants Introduced into North America. *Conservation Biology* 11: 193–203.

Tiedje, J.M., R.K. Colwell, Y.L. Grossman, R.E. Hodson, R.E. Lenski, R.N. Mack, and P.J. Regal. 1989. The Planned Introduction of Genetically Engineered Organisms: Ecological Considerations and Recommendations. *Ecology* 70: 298–315.

# 12

# Potential Impacts of Genetically Modified Trees on Biodiversity of Forestry Plantations
## *A Global Perspective*

BRIAN JOHNSON AND KEITH KIRBY

Plantations have been a major part of European forestry for at least 300 years and are increasingly important in commercial forestry throughout the globe. Research has revealed much about how the composition and management of plantations affect the ecology of the natural biodiversity in and around them. The ecological principles revealed by this research effort, which is centered in the conifer plantations of Europe and the pine plantations of the southern United States, are applicable to plantations worldwide.

Contrary to popular belief, many plantations harbor a significant part of natural woodland biodiversity, and some are crucial to the survival of certain specialized birds and mammals, especially in areas where plantations are managed specifically to enhance wildlife. There may be compelling environmental reasons for increasing plantation areas in the future, including introducing genetically modified (GM) plantations, if adequate biosafety can be achieved. Commercial pressure on old-growth forest could be alleviated if new plantations replaced farmland or low-density plantations, with considerable benefit to global biodiversity. Improvements to forest trees via conventional selective breeding are severely limited, but transgenic technology offers the opportunity to domesticate trees to tailor their characteristics more closely to the requirements of commercial forestry and the end user of forest products.

Transgenic techniques can produce varieties that enable different management regimes to be used to grow them. This is especially important in agricultural crops and trees, where agrochemical use can be affected by transgenic traits such as herbicide tolerance and pest and disease resistance. Such changes can have both adverse and beneficial impacts on native biodiversity in contact with agriculture and forestry. For example, the potential impact on biodiversity of transgenic herbicide-tolerant (GMHT) forest trees lies in their ability to withstand the application of broad-spectrum herbicides, which are used to control competing vegetation, especially in the early stages of plantation establishment. The early stages of plantations are known to be important for woodland biodiversity, whether plantations are being newly established on open land or are replacing felled trees. It might be expected that the use of herbicides on GMHT trees would lead to widespread weed kill, which in turn would give improved tree growth and quicker and more complete canopy closure. If herbicide-tolerant trees are ever to be used in GM plantations, there would need to be management schemes to counter such adverse effects on natural biodiversity. Because weed control should be easier and more certain of success in GMHT forest stands, managers might be more inclined to leave large areas unsprayed to act as the equivalent of biodiversity-rich "rides" and "glades" in non-GM forests.

Changing the quality of timber in growing trees has long been a goal of plant breeders. Transgenic technology offers a way of achieving radical changes in the lignin-to-cellulose ratios of both conifers and deciduous hardwoods. Doing that, however, has the potential inadvertently to alter the susceptibility of such trees to animal and fungal attack, with consequent damage to the trees. This could call for either more agrochemical management of such plantations or the insertion of multiple traits such as insect, fungal, and viral resistance to combat damage. If such traits were to be transferred inadvertently to wild-type trees, they would have the potential to increase fitness, perhaps leading in the long term to ecosystem dysfunction.

Concerns about gene transfer to and from transgenic trees have led to increased research into mechanisms for engineering sexual incompetence into transgenic trees. The result could be trees that produce little pollen and few flowers, seeds, and fruits. Not only are these food resources important for supporting biodiversity in coniferous and deciduous hardwood forests, but they are also economically valuable in some parts of the world. If sexual incompetence becomes necessary for biosafety reasons, then single-variety plantations may be undesirable, and mixed-variety (GM and/or conventional) stands would be preferable. There may also be biosafety advantages in growing mixed stands of GM and conventional trees, especially in reducing selection pressures for the evolution of pest resistance. Mixed stands could also be designed to further reduce gene flow in situations where sexual incompetence mechanisms in GM trees are not totally effective.

Assessing risks from trees derived from transgenic methods will be complex and challenging. In assessing risks to biodiversity from annual transgenic agricultural crops, it is possible to carry out comparative ecological studies, such as the farm-scale evaluations of GMHT crops in the United Kingdom. With long-lived perennials such as trees, however, a predictive modeling approach is the only realistic option if regulatory authorities are to make decisions about release within a reasonable period. Debate about the use of biotechnology to create transgenic plants and animals has largely centered on the perceived direct risks and benefits of introducing new genes into crops, fish, and trees. Concerns about food safety and gene transfer have dominated the debate, and there has been relatively little discussion about how plants and animals possessing transgenic traits might make possible new forms of agriculture, aquaculture, and forestry management, with consequent indirect risks and benefits. Historically the impacts of new management regimes in agriculture, fish farming, and forestry have been profound, changing whole landscapes and their associated biodiversity. In this chapter we focus on assessing the possible effects on biodiversity of the management of forest plantations of genetically modified trees. We do not address in detail ways in which the use of transgenic trees might influence wider forestry strategy, such as the possibility that high-yielding GM forests could replace old-growth cropping as a prime source of high-quality timber, pulpwood, and other forest products.

## Plantation Ecology

Large-scale afforestation of agricultural and marginal land has been practiced in Europe since the eighteenth century, and in some areas old-growth forests have been replanted either in part or by complete replacement of the ancient trees. Plantations now make up over 50% of all forests and woodlands in Europe and supply a higher proportion of timber and other forest products than old-growth forests (Peterken 1993). It is estimated that the area of plantations worldwide is now over 185 million hectares, of which around 60% is in temperate regions and around 40% in the tropics (FAO 2000).

Plantations are not only the most productive and profitable forest areas; they can also make a significant contribution to general biodiversity. In some regions of Europe they harbor most of the forest biodiversity. However in other parts of the world, plantations make a relatively minor contribution to forest biodiversity, often replacing highly diverse ecosystems with species-poor monocultures. Some tropical plantations on low-grade farmland can be ecologically diverse. For example, in Malaysian plantations of *Eucalyptus,* moth diversities can be as high as in natural secondary forest (Chey et al. 1997).

In Europe and North America, the planting dates and management history of new and replacement woodlands are sometimes well documented, allowing research to be carried out comparing plantations of different ages and tree species in a large number of different locations with differing management regimes. The biodiversity of American and European plantations has therefore been the subject of much research over the past 40 years, with the results being used as a basis for planting schemes and management regimes that favor biodiversity in plantations that also give high timber-crop yields.

In North America, Asia, and parts of Central and Eastern Europe, plantations on previously afforested areas are increasingly replacing old-growth forests that have been logged, leading to concern about the survival of species associated with ancient forest areas, some of which are still being logged. Conservationists and foresters have argued that to maintain timber production in these regions, substantial areas of new plantations on open land will be necessary to alleviate commercial pressure on natural forests, especially those that still have large areas of primary, old-growth forest ecosystems (Shepherd 1993).

These same arguments are increasingly being applied to tropical areas of Africa, Asia, and the Pacific Rim, where commercial pressure on rain forests and other natural ecosystems is leading to severe degradation of natural biodiversity, climate, water resources, and soils. Plantations of commercially valuable trees such as teak, mahogany, and eucalyptus are increasing in area, as logging from natural forests becomes more difficult both physically and politically.

If plantations are to replace a significant proportion of present-day commercial production from old-growth forests, there is a strong argument to make them as productive as possible in both yield and quality and also to lower management inputs. This "intensive silviculture" (Moore and Allen 1999) is rapidly being adopted throughout the world as a way of increasing production per unit area while simultaneously reducing unit production costs. In common with the development of transgenic agricultural crops, most research on GM trees has concentrated on traits to help secure these goals. Pest and disease resistance, modified quality traits such as altered lignin content, and tolerance to herbicides have been the main traits for transgenic trees, with current research also focusing on domestication of trees through transgenic traits for accelerated breeding (e.g., precocious flowering) and genomics research. There is also increasing research interest in genetically modifying trees to enable them to be grown in saline and arid soils.

There has been little consideration of the effects that plantations of GM trees might have on forest ecology and biodiversity. Generally, however, in intensively managed landscapes, the establishment of plantations on land that has been previously managed as arable farmland and sown grassland results in a net gain for biodiversity (Nature Conservancy Council 1991). On the other hand, replacement of old-growth forest and seminatural vegetation by plantations often lowers biodiversity, favoring colonizing, ruderal flora and fauna over the diverse wildlife of natural ecosystems (Nature Conservancy Council 1986), although that is not always the case for all taxa. For example, recent research in the United Kingdom reveals that species richness of macrofungi in conifer plantations is as high as that in seminatural pine and oak woodlands (Humphrey et al. 2000).

The biodiversity of forest plantations depends on a number of factors, of which the following are the most important.

- *Plantation size.* Larger plantations generally support more biodiversity (Peterken 1993). This is mainly because in large forests there are likely to be more sources from which propagules of woodland species can colonize the plantations, and larger forests are more likely than small woodlands to contain a wide variety of geomorphological features. This means that larger plantations are likely to contain more biological and physical "niches" that woodland wildlife can exploit.

- *Sources of woodland species already on site or nearby.* Peterken and Game (1984), in an extensive review of the biodiversity of plantations in the United Kingdom, found that woods that had been planted adjacent to ancient woodland and old woodland/hedge features were much richer in woodland biodiversity, and so were those that included water features such as streams and rivers.

- *Nutrient levels and geology of soils.* High nutrient levels on clay soils give an impoverished ground flora composed of a few very competitive species, whereas the ground flora of plantations on sands, peat, and loams with lower nutrient levels are generally more diverse (Ferris et al. 2000). Soil types also affect the capacity of plantations to acidify watercourses within them. Peat soils in particular can exacerbate acidification, adversely affecting the ecology of forest streams and the lakes and rivers into which they flow (Stoner et al. 1984). The replacement of deciduous forest by conifers can also significantly reduce stream flow by increasing transpiration and intercepting rainfall, damaging the ecology of woodland streams (Swank and Douglass 1974).

- *Age of forest.* Studies comparing vascular plant colonization of plantations of different ages (e.g., Peterken 1993) have found that older plantations are slightly richer than recently planted stands, but after around a century of establishment, plantations do not gain in vascular plant richness, presumably because as new species colonize, early colonizers die out. Although rates of colonization of specialist woodland plants are often very slow, especially into closed-canopy, mature stands, some insects, birds, and mammals rapidly colonize new stands, taking advantage of the physical and biological resources that the trees themselves provide.

- *Tree species.* In the temperate zone, insect and bird diversity is generally much greater in deciduous stands, especially those where abundant tree flowers and seeds are available. Deciduous stands usually have a greater diversity of physical structure than conifers, especially if the latter are even-aged, single-species stands (Moss 1979). Tree species diversity within plantations is also important because stands that have mixed tree species can be expected to have a wide range of tree architecture and food resources, not only from the foliage, flowers, and seeds of living trees, but also in the range of saprophytic fungi and other species feeding on dead-wood.

- *Management.* Management is probably the most important factor influencing the level of biodiversity found in forests. Biodiversity is greater in stands that are managed for diverse forest architecture, especially where areas of old-growth forest are maintained as "reservoirs" of native species. In both plantations and native forests, small coup (i.e., harvest) areas and short rotations support higher natural biodiversity because they result in variable age structure between coups and more regular opening up of the forest floor, allowing flora and fauna to benefit. Small coup areas also provide more forest rides and firebreaks, which are also very important both as sources of colonization and for the provision of "forest-edge" habitats, which favor high insect and bird diversity. Studies in the United States and Europe (Mitchell and Kirby 1989) indicate that the more diverse the spatial and temporal physical architecture, the greater the biodiversity of the forest, no matter which tree species are present or planted.

Generally, high plant abundance and diversity of "forest weeds" in early stages of plantation growth are important for the survival of insects and breeding birds (Moss 1979). The highest natural biodiversity is found where managers carry out localized weed control around the growing trees, leaving areas between coups to undergo natural succession. However, in many parts of the world, tolerating weeds (including native trees and shrubs) in young plantations could have a severely damaging effect on establishment of the planted trees, especially in some parts of the tropics, where colonization by grasses and native woodland shrubs can easily outcompete the crop. This may be less of a problem in temperate regions, where tree growth can overcome competing flora, often with little intervention other than preplanting weed control. Single-application weed control measures may not have long-lasting effects on noncrop plant and insect diversity, and if forest managers use herbicides carefully, leaving some areas untreated, biodiversity can be maintained (Morrison and Meslow 1984; Santillo et al. 1989; Mellin 1995).

The sporadic use of insecticides in forestry management can have adverse impacts on nontarget animals, although the effects are usually confined to short periods after application, with recolonization rapidly replacing the fauna. Stribling and Smith (1987) showed that in oak/maple forests in the United States, spraying insecticides for the control of gypsy moth did not appear to damage bird populations. In contrast, pine beauty moth control programs in Canada had short-term adverse effects on birds (Spray et al. 1987).

# Potential Effects of GM Trees on Plantation Biodiversity

Biotechnology is an important extension of conventional breeding techniques, not least because the technology enables radical changes in phenotypes through the insertion of gene cassettes that confer new traits. These changes are often not achievable using conventional breeding techniques, partly because the time required to find or construct suitable genes would be inordinately long, but mostly because the genes necessary to produce the desired traits are not present in the gene pool of the species and its ancestors.

Transgenic techniques, like other forms of plant breeding, produce varieties that exhibit different phenotypes from their ancestral form, which in turn enable different management regimes to be used to grow them. This is especially important in agricultural crops and trees because transgenic traits such as herbicide tolerance and pest and disease resistance can change the use of agrochemicals, with both adverse and beneficial impacts on native biodiversity in contact with agriculture and forestry (Johnson 2000). Conventional breeding can produce agricultural crops with similar traits (e.g., herbicide tolerance and pest resistance), but selecting for such traits is very difficult in forestry tree species, where plant breeders have to contend with long generation times and traits that may manifest themselves only at maturity. With the advent of biotechnology and cloning techniques comes the real possibility of developing domesticated trees that produce timber and other products that are closer to market needs and which can be grown more easily and quickly in plantations.

The impact of such trees on biodiversity is likely to depend much more on the traits they possess and the way in which plantations are managed than on the process by which such traits are achieved. Currently the traits discussed below are more easily achieved using transgenic techniques, but in the near future, increased knowledge of tree genomics, coupled with marker-assisted breeding, may produce similar results.

## *GM Herbicide Tolerance*

As we show in Table 12-1, herbicide tolerance (GMHT) is one of the main areas of research and development in transgenic trees. This research effort has focused on achieving tolerance to broad-spectrum herbicides such as glyphosate and glufosinate-ammonium, which are commonly used in conventional forestry to control vigorous species of grasses, herbs, and shrubs that compete with the planted crop.

The main effect of using herbicides in forestry plantations, whether conventional or GMHT, is the potential for destruction of native woodland flora and dependent fauna already present. Drift to watercourses and to marginal habitats such as hedge banks can result in destruction of flora that are poten-

**TABLE 12-1.** Number of GM Tree Field Trials, 1988–2000

| Genus | Herbicide tolerance | Insect resistance | Virus resistance | Lignin | Markers | Other | Total |
|---|---|---|---|---|---|---|---|
| Betula | | | | 1 | 1 | | 2 |
| Castanea | 1 | | | | | | 1 |
| Corcia | | 10 | | | 2 | | 12 |
| Eucalyptus | 4 | | 1 | 2 | 3 | 2 | 12 |
| Juglans | | 7 | 1 | | | 7 | 15 |
| Liquidambar | 3 | | | | | | 3 |
| Malus | 5 | 6 | | | 2 | 16 | 29 |
| Olea | | | | | 2 | 2 | 4 |
| Picea | | 3 | | | | 3 | 6 |
| Pinus | 1 | 1 | | | 11 | 2 | 15 |
| Populus | 41 | 36 | | 10 | 14 | 42 | 143 |
| Prunus | | | 3 | | 3 | 1 | 7 |
| Pyrus | | | | | | 3 | 3 |
| Total | 55 | 53 | 15 | 13 | 41 | 75 | 252 |

*Note:.* "Other" traits include fungal resistance, salt tolerance, altered flowering, and faster growth.

*Source:* Rautner, M. 2001. Designer Trees. *Biotechnology and Development Monitor* 44: 2–7.

tial colonizers of the developing woodland. Preplanting destruction of these flora and their associated fauna could result in impoverished forest ecosystems in GMHT plantations in later years. Preplanting herbicide regimes for GMHT plantations would be similar to those already being used and can be expected to have similar impact on native biodiversity. At present, herbicide application after planting often risks damaging conventional trees, especially where competing wild plants require high concentrations of herbicides for effective control. The introduction of herbicide tolerance would allow broad-spectrum weed control at any stage in plantation development. It would most likely be used, however, in the early stages of tree growth, when a wide range of forest and forest-edge species tend to be present.

If GM trees tolerant to broad-spectrum herbicides were to be widely used, GMHT plantations could be less attractive for species of birds and invertebrates that rely on young plantation habitat, with its combination of young planted trees and diverse wild plants that support the food webs on which they rely. Santillo et al. (1989) and Morrison and Meslow (1984), for example, found that herbicide treatments of felled and replanted forest significantly reduced the abundance and diversity of phytophagous arthropods. Studies in Sweden have linked the diversity and abundance of insectivorous birds with the availability of herbivorous insect larvae in *Picea* forests (Atlegrim and Sjöberg 1996).

Widespread and routine application of broad-spectrum herbicides to new plantations might prevent the establishment of woodland understory species that rely for germination on the light and moisture of newly planted stands. Woody shrubs generally cannot germinate in closed-canopy forests; they establish themselves either early in the plantation successional stages or only very much later after the stand has been thinned. In the absence of these shrubs, plantations are poorer in wildlife because they are not only simpler in species composition but also in terms of structural diversity.

### Tolerance to Adverse Soils

Drought resistance and tolerance to acidic and saline soils are active areas of research into transgenic trees. Achieving these characteristics may eventually permit afforestation of areas having soils in which commercially valuable trees currently cannot grow successfully. This could be beneficial to biodiversity and general environmental health if it allowed trees to thrive on agriculturally degraded soils, providing soil stability, soil refurbishment, and carbon sequestration.

However these developments also raise the possibility of plantations on soils that do not naturally support a characteristic woodland flora and fauna. Plantations on saline soils may develop a ground flora and understory of species found in shaded gullies in saline habitats such as salt flats and salt marshes. In the tropics, salt-tolerant plantations might support associated wildlife characteristic of mangrove swamps, especially in coastal and river floodplain locations. However, there may be relatively few plant species capable of tolerating the double stress of shade plus high salt levels.

The implications of these developments for biodiversity are difficult to predict from ecological theory. Only by establishing pilot-scale plantations can ecologists begin to understand the impacts of afforesting saline and arid areas, with their established and highly adapted ecosystems.

# Pest Resistance

Some contributors to the GM debate have argued that the introduction of insect resistance into forest trees would render plantations devoid of phytophagous and pollen-feeding insects (Owusu 1999; Tickell 1999). Although that may be a risk where GM trees produce broad-spectrum anti-feedants and insecticides (such as lectins and protease inhibitors), more specific genetically modified insect resistance (GMIR) such as *Bt* is only likely to have significant adverse effects on woodland ecosystems where the trait affects keystone phytophagous species, such as Lepidoptera and Coleoptera, that are themselves endangered or are crucial parts of food webs supporting rare or endangered species of birds or mammals. This could be a greater risk in tropical areas, where rare species may be more specialized in their food and habitat requirements than those in temperate zones. The introduction of trees containing *Bt* toxins in these situations could have a significant adverse effect on these rich ecosystems. Lectins and protease inhibitors may be more generalized in their effects and risk destroying or deterring most arthropods, as well as mammals and birds feeding on the trees, but exposure to the toxins would depend on levels of gene expression in various parts of the GM trees.

These risks must be assessed in relation to the risks posed by the use of insecticides to control pests in conventional plantation management. Risks from conventional pesticide use are relatively low due to the sporadic nature of insecticide use in conventional forestry and the capacity of woodland fauna to recolonize after such applications. Without comparative research it is difficult to estimate relative risks to biodiversity from introducing GMIR traits into trees. The problem is compounded by the difficulty of predicting long-term effects, especially where both target and nontarget insects could develop resistance to the trait. Current proposals for managing such risks usually assume that resistance in insects would effectively be a recessive trait if high doses of *Bt* toxin were used in the transgenic plants (Andow et al. 1998). This is a rather speculative assumption that, if incorrect, could prove fatal to resistance management schemes relying on maintaining refugia of susceptible insects (Gould 1998; Andow et al. 1998). If pest populations and nontarget insects develop resistance to the GMIR trait—and that is likely if the trait is inserted into long-lived trees—then there could be a long-term management trend toward increased insecticide use. Although introduction of an IR trait may initially reduce target pest populations, it could allow other, previously rare "secondary" pest species to flourish, leading to increased need for conventional chemical control (Sharma and Ortiz 2000). Ashouri et al. (2001) have shown that GMIR potatoes designed to resist Colorado potato beetle (*Leptinotarsa decemlineata*) had unexpected effects on populations of potato aphids feeding on the transgenic plants. In one transgenic line they studied, those effects could have had a tertiary impact on the spread of potato viruses. This work shows that even in a simple agricultural ecosystem it is difficult to predict ecological perturbations caused by GM plants with simple monogenic insect resistance.

IR traits might affect the soil and wood decomposer cycle within plantations because the initial stages of decomposition in forest ecosystems often involve comminution of plant material by arthropods, especially mites (Edlin 1970). If the GM trait were adversely to affect this initial stage of decomposition, then leaf litter and brash could break down more slowly, with unpredictable ecological consequences for arthropod and fungal components of the forest ecosystem.

## Disease Resistance

Traits conferring general fungal and viral resistance have been inserted into tree species, mostly into fruit-producing varieties such as papaya, apple, and cherries (Table 12-1). Although virus resistance is unlikely to have an adverse impact on the biodiversity of plantations, it is possible that the introduction of generalized fungal resistance could affect decomposer ecosystems in plantations, although there are a large number of fungal species involved in such processes in woodlands and it is likely that the traits would be overcome by at least some of the fungi present. However many saprophytic fungi are quite specialized in their choice of substrate, so there could be adverse effects on the diversity of macrofungi in GM fungal-resistant plantations. It is also possible that antifungal traits could affect the important and complex relationships between ectomycorrhizal fungi and tree roots.

Concern about the conservation of fungi has increased in recent years because of a significant decline in the abundance of many fungal species throughout Europe. Humphrey et al. (2000) found that mature stands of pine and spruce held the greatest diversity of fungi and that clear felling with removal of deadwood was associated with significant reductions in fungal species diversity. Of the 419 species they recorded in plantations, 157 were litter saprophytes. As the management of GM plantations might differ from that of conventional stands, the impact on fungal diversity may become an important aspect of monitoring.

## Quality Traits

So far most research on changing the quality of timber has focused on varying lignin/cellulose ratios in deciduous woods such as poplars (*Populus* spp.). Wood from such genetically modified trees would be better suited to the industrial processes used to recover cellulose for paper and board manufacture, potentially leading to significant reductions in the chemicals and energy needed for pulping. That in turn should reduce pollution from mills (Petit-Conil et al. 2000; Pilate et al. 2002).

Little is known about possible effects of altering lignin and cellulose content on palatability of GM trees to phytophagous animals, although reduced-lignin poplars released experimentally appear to be no more susceptible to insect attack than conventional poplars. Other trees with altered quality traits may, however, be found to behave as some agricultural crops do, requiring more defense against phytophagous arthropods and fungus and virus attack. If such defense were to be in the form of increased agrochemical use, then the biodiversity value of the GM plantations might be lower than that of conventional plantations. Alternatively, it might be necessary to introduce multiple GM traits, such as insect, fungal, and viral resistance, to combat damage.

Changing lignin-to-cellulose ratios may affect the strength characteristics of trees, which could affect their wind firmness in plantations. That could lead to changes in how and where they are grown or in their patterns of response to extreme wind conditions. Such changes could influence forest strategy, for instance by encouraging more plantations in less-exposed situations. There may be implications for biodiversity in such changes.

## Sterility and Other Biosafety Traits

Genetic modification is often used to manipulate tree species that are either native to the region in which they are to be grown or sexually compatible with native trees that are closely related. Trees are by nature long-lived perennials, and regulatory authorities are rightly concerned about genetic stability and gene flow from the transgenic varieties to native species. In some species, notably poplars and aspens (*Populus* spp.) and willows (*Salix* spp.), there is concern that transgenics may be able to propagate vegetatively, but most risk assessment has centered on gene transfer via pollen.

There has been a trend toward trying to engineer sexual incompetence into transgenic trees, either by disrupting pollen production mechanisms or by suppressing flowering and therefore the production of seeds and fruits. If these measures were adopted commercially, plantation trees would produce neither pollen nor fertile ovaries. A major element of the woodland food web could be either unavailable or greatly reduced in such plantations.

The production of pollen, nectar, seeds, and fruits is an important factor in maintaining natural biodiversity in uniform, single-species plantations of conventional trees, whether coniferous or deciduous (Palik and Engstrom 1999). In temperate regions, for example, coniferous plantations are crucial to the survival of several species of seed-eating birds, such as crossbills in Europe and Clark's nutcracker in the United States, and some mammals, such as red squirrels in Britain and the red tree-mouse in the United States. Specialist birds, such as the yellow-bellied sapsucker, depend on a range of conifer and deciduous tree saps and nectars at critical points in their breeding cycles (Tate 1973). In tropical regions, deciduous plantations often produce copious quantities of nectar and pollen that support a wide range of insects and the food webs associated with them. Honeybees often feed on deciduous hardwood plantation species and in some areas provide a valuable source of income. There is therefore a conflict between the desire to enhance biosafety by suppressing sexual reproduction in transgenic trees and the need to maintain biodiversity and, in some areas, economically important activities such as plantation-based beekeeping.

The most obvious solution to maintaining biodiversity among sterile trees would be to plant mixed stands of transgenic and conventional trees—either direct admixtures or blocks of different trees. Mixed-species stands are known to increase forest biodiversity by increasing structural complexity, but as Larsen points out, tree species for admixture should be chosen to support food web interactions by taking account of known coevolutionary relationships. Plantations of this type should be more resistant to physical or biotic stress (Larsen 1995).

Because genetically engineered sexual incompetence will occasionally fail, there could be biosafety implications for mixed-species plantations if the species admixture included trees that were sexually compatible with transgenics. If the mixed species were sexually incompatible, risks could be acceptably low, and such mixtures could add to biosafety by further reducing vegetative propagation and pollen flow from GM trees that sporadically regain fertility.

## Growth Traits

Transgenic traits to produce faster growth have been suggested as a possible solution to improving the productivity of biomass plantations and those where pulpwood is the primary goal. Such plantations could be advantageous to native biodiversity, especially if they were subjected to regular short-rotation coppicing favoring woodland flora and bird species that require a range of ecological successions within forests.

Even if biomass and pulpwood coppicing with regular fertilizer use were to result in species-poor ground flora, the temporal and spatial range of vertical structure produced could favor woodland flora and fauna, including bird species that require a range of structural successions within forests. However, these potential advantages could be offset by simplification of plantation food webs caused by high-growth GM trees that included pest- and disease-resistance traits.

## Risk Assessment

There are serious concerns about introducing high-growth traits and pest and disease resistance into transgenic trees, especially where the tree species already shows invasive tendencies. There are examples in which nontransgenic conifers and deciduous trees have invaded natural ecosystems (for example, *Hippophae rhamnoides, Pinus sylvatica, Rhododendron ponticum, Robinia pseudacacia,* and *Quercus ilex* in Europe, and *Syringa* in the United States). These invasions have often been from plantations into surrounding natural woodland. Research has shown that predicting invasiveness is very difficult, especially for long-lived perennial species such as trees (Williamson and Fitter 1996; Manchester and Bullock 2000).

Risk assessment processes worldwide have not yet been able to predict with any certainty what impact releasing transgenic perennial plants such as trees might have on native biodiversity. The issue of gene flow has dominated the debate, which has largely focused on attempts to measure *rates* of gene transfer (the question, Does it happen?), with little research on the *impacts* that transgenes might have on fitness and other ecological characteristics of recipient organisms (the question, Does it matter?). The latter is especially difficult to predict in the longer term where long-lived perennials such as trees are involved. The key to predicting fitness lies in identifying the principal components of fitness relative to the organism in question and the habitats within which it lives. Muir and Howard (2001) have modeled impacts on fitness of a theoretical transgene that influences several fitness parameters simultaneously in a fish, Japanese medaka (*Oryzias latipes*). They estimate that for a wide range of fitness values, transgenes conferring quite small increases in fitness could spread in medaka populations. Using a similar methodology, models could be developed for estimating impacts of transgenes (such as increased growth) on tree fitness, provided that there are sufficient data available for identifying and estimating fitness components. A major problem with the modeling approach to fitness assessment of transgenic perennials lies in the difficulty of predicting the environments into which such trees might spread. Unless that can be done with some certainty, it will not be possible to estimate fitness components to populate a model.

If GM trees are ever to be used in plantation forests, it is likely that they will contain combined GM traits, such as altered timber quality with insect resistance. It could be very difficult to model the direct and indirect impacts on biodiversity of these GM multiple-trait phenotypes, even if the potential ecological consequences of each trait in isolation were known. In assessing risks from annual GM agricultural crops, it is possible to carry out comparative ecological studies over three or four years, as in the farm-scale evaluations of GMHT crops in the United Kingdom (Firbank et al. 1999; Firbank and Forcella 2000). With long-lived perennials such as trees, however, a predictive modeling approach may be the only realistic option if regulatory authorities are to decide applications for release within a reasonable time. Where the ecological relationships between trees and target conservation species are known, it may be possible to model impacts. An example is the use of hybrid poplars by breeding golden orioles (*Oriolus oriolus*) in the United Kingdom, where the physical structure and phenology of the trees are crucial to the bird's breeding success. The critical factors in this ecological relationship are known (Milwright 1998), so that transgenic poplars could be assessed for their suitability as golden oriole habitat.

The use of ecological modeling may improve the ability of risk assessors to predict biodiversity impacts, but comparison between transgenic and conventional plantations can only be made if models can be populated with quantitative data. To provide adequate data for risk assessment requires more research on the population dynamics of transgenic trees and the general ecology of plantations, especially forests populated by tree species that are currently the focus of transgenic and genomic development. There is a particular need for more research into the ecology of plantations in the tropics, where the relationship between forest trees and non-crop biodiversity is important for nature conservation and may be critical to the stability and productivity of future plantations, whether transgenic or conventional (Larsen 1995).

## Conclusions

Regulatory authorities, as well as organizations responsible for forestry strategy and management, will want to understand the characteristics of transgenic trees before growers put long-term investment into GM plantations. They will especially want to understand the risks to forest ecosystems, and they would probably prefer to use evidence from pilot-scale studies rather than predictive modeling for risk assessment, although modeling may be the only option available if the long-term impacts of traits such as pest and disease resistance are to be better understood in the near future. Without evidence from field testing and pilot-scale studies, it is difficult to see how risk can be assessed in a way that is both scientifically defensible and publicly acceptable. Traits such as herbicide tolerance and sterility are more easily field tested for ecological effects in the shorter term, and pilot-scale studies should lead to an understanding of whether and how potential negative impacts can be minimized by risk management measures. Such studies may also help forest managers determine whether herbicide tolerance could be used to increase biodiversity in the early stages of plantation establishment, for example by allowing herbicides to be applied later than they currently are.

Before any pilot-scale studies can begin, regulatory authorities first need to decide whether some traits should be released at all, especially for insertion into native tree species. This risk assessment needs to be applied as much to mutational and marker-assisted breeding and other "conventional" techniques as to transgenic techniques. Because there can never be a guarantee of zero gene flow between sexually compatible individuals, there are strong arguments for a precautionary approach to genetic transformation of native plant species, especially where traits that could increase fitness are concerned. It would be wise to confine the first pilot-scale studies of transgenic trees expressing traits such as pest and disease resistance to tree species that are not native to the planting location and to engineer sexual incompetence into such trees. Setting up well-designed and relatively safe field-scale studies comparing the ecological impacts of transgenic trees (and the management systems associated with them) with those of conventional plantations should be a first step in the proper assessment of risks from commercial-scale plantations.

## Acknowledgment

The authors thank Anna Hope, English Nature, for reviewing and providing helpful input into an early draft of the manuscript.

## References

Andow, D.A., D.N. Alstad, Y.-H. Pang, P.D. Bolin, and W.D. Hutchison. 1998. Using an F2 Screen to Search for Resistance Alleles to *Bacillus thuringiensis* Toxin in European Corn Borer (Lepidoptera: Crambidae). *Journal of Economic Entomology* 91: 579–584.

Ashouri, A., D. Michaud, and C. Cloutier. 2001. Unexpected Effects of Different Potato Resistance Factors to the Colorado Potato Beetle (Coleoptera:Chrysomelidae) on the Potato Aphid (Homoptera:Aphididae). *Environmental Entomology* 30 (3): 524–532.

Atlegrim, O., and K. Sjöberg. 1996. Effects of Clear-Cutting and Single-Tree Selection Harvests on Herbivorous Insect Larvae Feeding on Bilberry (*Vaccinium myrtilis*) in Even-Aged Boreal *Picea abies* Forests. *Forest Ecology and Management* 87: 139–148.

Chey, V.K., J.D. Holloway, and M.R. Speight. 1997. Diversity of Moths in Forest Plantations and Natural Forests in Sabah. *Bulletin of Entomological Research* 87(4): 371–385.

Edlin, H.L. 1970. *Trees, Woods and Man*. London: Collins.

FAO (UN Food and Agricultural Organization). 2000. *Forest Resources Assessment 2000*. www.fao.org/forestry/fo/fra/index.jsp (accessed November 29, 2002).

Ferris, R., A.J. Peace, J.W. Humphrey, and A.C. Broome. 2000. Relationships between Vegetation, Site Type and Stand Structure in Coniferous Plantations in Britain. *Forest Ecology and Management* 136(1–3): 35–51.

Firbank, L.G., A.M. Dewar, M.O. Hill, M.J. May, J.N. Perry, P. Rothery, G. Squire, and I.P. Woiwod. 1999. Farm Scale Evaluation of GM Crops Explained. *Nature* 339: 727–728.

Firbank, L.G., and F. Forcella. 2000. Genetically Modified Crops and Farmland Biodiversity. *Science* 289: 1481–1482.

Gould, F. 1998. Sustainability of Transgenic Insecticidal Cultivars: Integrating Pest Genetics and Ecology. *Annual Review of Entomology* 43: 701–726.

Humphrey, J.W., A.C. Newton, and A.J. Peace. 2000. The Importance of Plantations in Northern Britain as a Habitat for Native Fungi. *Biological Conservation* 96(2): 241–252.

Johnson, B.R. 2000. Genetically Modified Crops and Other Organisms: Implications for Agricultural Sustainability and Biodiversity. In *Agricultural Biotechnology and* the Poor: Proceedings of an International Conference, Washington, DC, 21–22 October 1999, edited by G.J. Persley and M.M. Lantin. Washington, DC: Consultative Group on International Agricultural Research, 131–138.

Larsen, J.B. 1995. Ecological Stability of Forests and Sustainable Silviculture. *Forest Ecology and Management* 73: 85–96.

Manchester, S.J., and J.M. Bullock. 2000. The Impacts of Non-native Species on UK Biodiversity and the Effectiveness of Control. *Journal of Applied Ecology* 37: 845–864.

Mellin, T.C. 1995. The Effects of Intensive Forest Management Practices on the Natural Vegetative Communities of Loblolly Pine Plantations in North Carolina. Master's thesis. North Carolina State University.

Milwright, R.D.P. 1998. Breeding Biology of the Golden Oriole *Oriolus oriolus* in the Fenland Basin of Eastern England. *Bird Study* 45(3): 320–330.

Mitchell, P.L., and K.J. Kirby. 1989. *Ecological Effects of Forestry Practices in Long-Established Woodland and Their Implications for Nature Conservation.* Occasional Paper No 39. Oxford, UK: Oxford Forestry Institute.

Moore, S.E., and H.L. Allen. 1999. Plantation Forestry. In *Maintaining Biodiversity in Forest Ecosystems,* edited by M.L. Hunter. Cambridge, UK: Cambridge University Press, 400–433.

Morrison, M.L., and E.C. Meslow. 1984. Effects of the Herbicide Glyphosate on Bird Community Structure, Western Oregon. *Forest Science* 30: 95–106.

Moss, D. 1979. Even-Aged Plantations as a Habitat for Birds. In *The Ecology of Even-Aged Plantations,* edited by E.D. Ford, D.C. Malcolm, and J. Atterson. Cambridge, UK: Institute of Terrestrial Ecology.

Muir, W.M., and R.D. Howard. 2001. Fitness Components and Ecological Risk of Transgenic Release: A Model Using Japanese Medaka (*Oryzias latipes*). *American Naturalist* 158(1): 1–15.

Nature Conservancy Council. 1986. *Nature Conservation and Afforestation in Britain.* Peterborough, UK: Nature Conservancy Council.

———. 1991. *Nature Conservation and the New Lowland Forests.* Peterborough, UK: Nature Conservancy Council.

Owusu, R.A. 1999. *GM Technology in the Forest Sector—A Scoping Study for WWF.* Godalming, Surrey, UK: World Wide Fund for Nature, WWF-UK.

Palik, B., and R.T. Engstrom. 1999. Species Composition. In *Maintaining Biodiversity in Forest Ecosystems,* edited by M.L. Hunter. Cambridge, UK: Cambridge University Press, 63–94.

Peterken, G.F. 1993. Long-Term Floristic Development of Woodland on Former Agricultural Land in Lincolnshire, England. In *Ecological Effects of Afforestation,* edited by C. Watkins. Oxon, UK: CAB International, 31–45.

Peterken, G.F., and M. Game. 1984. Historical Factors Affecting the Number and Distribution of Vascular Plant Species in the Woodlands of Central Lincolnshire. *Journal of Ecology* 72: 155–182.

Petit-Conil, M., C. Chirat, E. Guiney, W. Schuch, C. Lapierre, B. Pollet, I. Mila, W. Boerjan, L. Jouanin, and G. Pilate. 2000. Designing Lignin Structure for Modifying Pulping Performances: Application to Poplar Trees. *TAPPI Pulping/Process and Product Quality Conference, Boston, USA; Proceedings (Session 12).* Paper 12–1.

Pilate, G., E. Guiney, K. Holt, M. Petit-Conil, C. Lapierre, J.-C. Leple, B. Pollett, I. Mila, E.A. Webster, H.G. Marstorp, D.W. Hopkins, L. Jouanin, W. Boerjan, W. Schuch, D. Cornu, and C. Halpin. 2002. Field and Pulping Performances of Transgenic Trees with Altered Lignification. *Nature Biotechnology* 20: 607-612.

Rautner, M. 2001. Designer Trees. *Biotechnology and Development Monitor* 44: 2–7.

Santillo, D.J., D.M. Leslie, and P.W. Brown. 1989. Responses of Small Mammals and Habitat to Glyphosate Application on Clearcuts. *Journal of Wildlife Management* 53: 164–172.

Sharma, H.C., and R. Ortiz. 2000. Transgenics, Pest Management, and the Environment. *Current Science* 79(4): 421–437.

Shepherd, K.R. 1993. Significance of Plantations in a Global Forestry Strategy. *Australian Forestry* 56(4): 237–335.

Spray, C.J., H.Q.P. Crick, and A.D.M. Hart. 1987. Effects of Aerial Applications of Fenitrothion on Bird Populations of a Scottish Pine Plantation. *Journal of Applied Ecology* 24: 29–47.

Stoner, J.H., A.S. Gee, and K.R. Wade. 1984. The Effects of Acidification on the Ecology of Streams in the Upper Tywi Catchment in West Wales. *Environmental Pollution* A 35: 125–157.

Stribling, H.L., and H.R. Smith. 1987. Effects of Dimilin on Diversity and Abundance of Forest Birds. *Northern Journal of Applied Forestry* 4: 37–38.

Swank, W.T., and J.E. Douglass. 1974. Streamflow Greatly Reduced by Converting Deciduous Hardwood Stands to Pine. *Science* 185: 857–859.

Tate, J. 1973. Methods and Annual Sequence of Foraging by the Sapsucker. *Auk* 90: 840–856.

Tickell, O. 1999. GM Trees—Flowerless Forests. *Forestry and British Timber,* September 1999, 36–39.

Williamson, M., and A. Fitter. 1996. The Characters of Successful Invaders. *Biological Conservation* 78: 163–170.

# 13

# Transgenic Resistance in Short-Rotation Plantation Trees

## Benefits, Risks, Integration with Multiple Tactics, and the Need to Balance the Scales

KENNETH F. RAFFA

Nearly 20 years have passed since the first report of transgenic expression in plants (Meeusen and Warren 1989). It has been more than 10 years since development of transgenic *Bt* resistance in hybrid poplar against tent caterpillars and gypsy moths (McCown et al. 1991) and seven years since the first field trials (Kleiner et al. 1995). Yet we still have only a limited ability to assess the long-term implications of genetically modified organisms for forest ecosystems.

I would like to suggest four reasons for keeping the issue of environmental safety of genetically engineered trees at the forefront: First, trees pose some inherently distinct challenges. More than agricultural crops, trees have generation times much longer than those of their major pests. Also, trees are both commercial plantings and components of native ecosystems. The same species grow adjacent to each other and provide multiple ecological functions and human resources. Second, the track record of agricultural transgenic plants is too short to allow full evaluation (Parker and Karieva 1996; van Emden 1999). Problems arising from injudiciously selected biological control agents, calendar application of pesticides, vertebrate predator elimination, and fire suppression often were not detectable until many decades after implementation. In general, the longer the time lag between implementation and the appearance of problems, the greater the difficulty in addressing them. Third, pest managers and producers of seed for agricultural crops have implemented specific practices aimed at preventing biotype evolution (Roush 1997), but such practices have not yet been developed for trees, reflecting both the greater emphasis placed on agriculture and the difficulties of working with trees. Fourth, government-sponsored research on risk assessment of genetically modified organisms (GMOs) has fostered a false sense of assurance. As anyone who vies for competitive grants can attest, a key to success is defining tight, narrowly focused experiments that can be conducted under controlled conditions over short time frames, using organisms that lend themselves to simple bioassays. Unfortunately, the key questions about environmental safety are long term and complex and operate across multiple trophic levels.

My discussion here focuses on transgenic resistance against insects in plantation trees. However I hope that some general principles can be applied to other types of transgenic properties. I also need to emphasize that considerations appropriate to poplars may not apply to self-regenerating (i.e., non-sterile) and insect-pollinated trees. Likewise, questions about philosophical values, economic equity, and how society should determine an acceptable level of risk are outside the scope of this chapter but are addressed elsewhere (e.g., Strauss et al. 2000).

## Potential and Risks of Genetic Engineering

The advantages provided by genetically engineered pest resistance are significant (Strauss et al. 1997). Insects and pathogens are major limitations to tree productivity (Canadian Forest Service 1994). Moreover, the conditions under which trees are grown commercially, usually even-aged, evenly spaced monocultures, make them more manageable for the producer but also much more susceptible to the insects that consume them. Breeding for resistance is far more difficult in trees than in annual crops, for obvious reasons of size and time. Moreover, trees confront complexes of insects, and it is difficult to select for resistance against all of them. It is especially difficult when one insect prefers the very group of phytochemicals that confers resistance against others. For example, some phenolics are consistently associated with both high resistance against Lepidoptera and high susceptibility to Coleoptera (Raffa et al. 1997).

Deploying pest-resistant trees may allow reduced use of insecticides. This has been demonstrated in some agricultural crops (Gabrielle and Siedow 1999), although it is not validated in all systems (Obrycki et al. 2001). Moreover, enhancing productivity on intensively managed lands may alleviate demands on the land and resource base, thus allowing other regions to be set aside for wilderness and biodiversity objectives. Transgenic capabilities may also help us respond more quickly to invasive species, which are an increasing threat associated with global commerce.

A number of potential risks have also been identified. They include evolution of insect biotypes resistant to the transgene, adverse direct and indirect

**TABLE 13-1.** Differences in Properties of Systems Studied by Molecular Biologists and Ecologists

| *Molecular studies* | *Ecological studies* |
| --- | --- |
| Closed systems | Open systems |
| Simplified to best of ability | Complex interactions |
| Isolated mechanisms | Limits not precisely defined |
| Problems often resolved through technical advance or improved insight into a specific mechanism | Problems often resolved through recognition of additional factors impacting unit of study |

effects on nontarget organisms, and alteration of ecosystem processes such as nutrient cycling and pollination (NRC 2002). Molecular biologists and ecologists agree on both the potential benefits and the risks of transgenic plants, but often differ in the magnitude they assign to them. My perception is that this reflects differences in the levels of biological organization at which different scientists work (Table 13-1). Ecology, by definition, involves the study of interactions. Ecologists deal with systems that are open, complex, and to some extent undefined. Molecular biologists deal with systems that are closed, controlled, and simplified as much as possible. There is an inherent tension between the reality of the field and the precision of the laboratory. When a molecular biologist encounters a difficult problem, success often comes through a technical advance or an improved insight into a specific mechanism. That insight is often achieved by removing extraneous noise from the system. In the case of ecologists, our ability to resolve a question is usually improved once we identify additional factors that are affecting our unit of study. We often overcome a lack of understanding when we recognize how to introduce complex feedback into what we previously thought was a simpler system.

## Approaches to Risk Assessment

Based on such different experiences, molecular biologists and ecologists tend to approach questions of risk assessment from divergent notions about burden of proof. Molecular biologists frequently ask, What are the specific risks that can be anticipated, based on expression of a specific product? This question reflects a supposition that it is unscientific to assume that a risk might exist unless it can be well characterized or quantified. In contrast, ecologists tend to ask, What is the evidence that specific information derived from a closed system over a short time interval can be extrapolated to multiple interactions in open systems over long time periods? This question reflects a supposition that it is unscientific to assume without proof that extrapolation across scales will yield consistent relationships, just as it would be unscientific to interpret experiments without testing whether Heisenberg's Uncertainty Principle was satisfied. To ecologists, deployment of GMOs is a classic scaling problem, and their expectations are influenced by experiences in which extrapolating across scales does not yield predictable relationships (Naveh and Lieberman 1994; Turner et al. 2001).

Illustrations of how complex, time-delayed feedback can yield qualitatively different outcomes than would be predicted from single-interaction models arise from a large and diverse set of experiences. Although examples from the physical and social sciences abound, many also come from biological systems. One example comes from sheep production, where yield can be reduced by pests such as coyotes. Range managers attempted to eradicate coyotes, but early attempts were not successful because the technology was inadequate to compensate for coyotes' avoidance mechanisms. Eventually trap and bait technologies improved, and coyote populations decreased as planned. Unexpectedly, sheep populations declined also. The problem was that biologists had underestimated the complexity of the system. Coyotes also eat jackrabbits, whose numbers soared after coyotes were eliminated. They devoured the grass and starved the sheep (Wagner and Stoddard 1972; Longland 1991). Biologists had correctly identified the pathway that they intended to interrupt (i.e., coyotes eat lambs) but had failed to account for additional pathways that exert important feedback (Figure 13-1). Consequences such as these are not uncommon. Similar experiences occurred with attempts to remove wolves so as to favor beaver fur production; with fire suppression; and with DDT applications against spruce budworm (Furniss and Carolin 1980; Romme and Knight 1981; Naiman et al. 1994; Pollach et al. 1995; Estes 1996; Turner et al. 1997; Baskin 1999; Crooks and Soulé 1999; Henke and Bryant 1999; Estes et al. 2001). The general lesson is that when you transfer a technology from closed, controlled conditions to open, complex systems, there are almost always unforeseen parameters that affect system behavior. It is important to recognize that none of these problems arose from direct interactions, and hence they would not have been predicted by the types of experiments that current, federally funded risk assessment programs usually support.

Resolving such dichotomies between the experiences and assumptions of biologists who work at the molecular and the ecological levels is one of the most important steps in both maximizing the benefits and minimizing the risks of genetic engineering. In the case of pest resistance, ecological understanding can be useful both in identifying risks inherent in extrapolating from closed to open systems and in integrating multiple sources of insect mortality to improve efficacy. Likewise, molecular methods will provide some of the most valuable tools for ameliorating risks that ecologists identify (Raffa 1989).

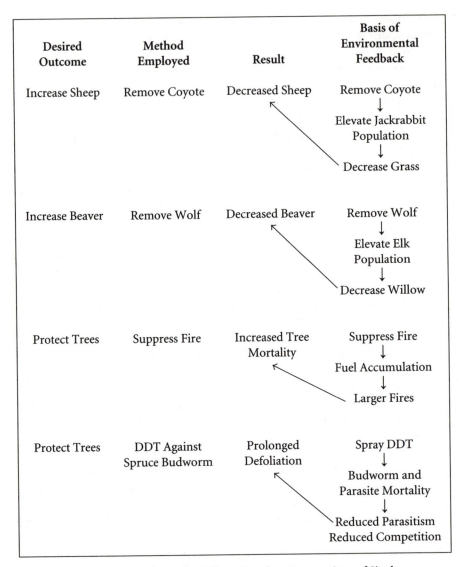

| Desired Outcome | Method Employed | Result | Basis of Environmental Feedback |
|---|---|---|---|
| Increase Sheep | Remove Coyote | Decreased Sheep | Remove Coyote ↓ Elevate Jackrabbit Population ↓ Decrease Grass |
| Increase Beaver | Remove Wolf | Decreased Beaver | Remove Wolf ↓ Elevate Elk Population ↓ Decrease Willow |
| Protect Trees | Suppress Fire | Increased Tree Mortality | Suppress Fire ↓ Fuel Accumulation ↓ Larger Fires |
| Protect Trees | DDT Against Spruce Budworm | Prolonged Defoliation | Spray DDT ↓ Budworm and Parasite Mortality ↓ Reduced Parasitism Reduced Competition |

**FIGURE 13-1.** Examples of Intended Effects Based on Assumptions of Single Interactions Generating Opposite Effects Once Opened to Multiple Interactions at Ecosystem Level

How can we best integrate molecular and ecological approaches to risk assessment? I suggest starting with a point of common agreement. Such a point is provided by the first National Academy of Sciences (NAS 1987) report on transgenic organisms, which has shown a remarkable ability to stand the test of time. The report emphasized that the "product not the process" should be the focus: "Assessment of risk should be based on the organism, not the method of engineering." Notably, a white paper by the Ecological Society of America independently reached a nearly identical and simultaneous conclusion: "[T]ransgenic organisms should be evaluated and regulated according to their biological properties, rather than according to the genetic techniques used to produce them" (Tiedje et al. 1989).

The product-not-process guideline is sometimes misinterpreted to mean that environmental concerns with GMOs are unwarranted. Nothing could be further from the truth. Rather, this principle dismisses scientifically unfounded concerns that a particular product may exert an effect simply because it was genetically engineered. But it also focuses attention on interactions. The question becomes, How will this organism interact with other organisms in complex, open systems, over long periods of time? Moreover, the NAS report yields a corollary: "If the product not the process is critical, then expertise in the methods of genetic engineering is not directly relevant to predicting how novel organisms will interact with ecosystems" (Raffa et al. 1997).

There is no uniform agreement on how to approach risk. One approach is to list every conceivable adverse consequence. That is perhaps not a bad starting point, but it loses scientific validity if it does not proceed to a rigorous, critical evaluation of those potential harms, or if it extrapolates uncritically from artificial conditions to ecosystem levels. Just as extrapolation from controlled to open systems can fail to detect risks, it can also exaggerate them. For example, laboratory assays clearly demonstrated that pollen from corn expressing *Bt* transgenes can kill monarch butterfly larvae (Losey et al. 1999). However subsequent studies showed the ecosystem-level impacts of "*Bt* pollen" to be minimal, localized to the immediate perimeter of cornfields, and far less damaging than the impact of habitat destruction (NRC 2002).

I see three disadvantages to the all-inclusive-lists approach. First, it can result in lost opportunities (NRC 2002). No undertaking is without risk, and refusal to manage it can preclude the realization of benefits such as those described above. For example, in cases where habitat loss is the most severe problem, might the increased productivity offered by transgenes improve opportunities for conservation and restoration? Ecology and economics must be recognized as integrated, not opposing, sciences; both terms arose from the same Greek root, *oikos* (Barrett and Farina 2000). There is value to recognizing that both deal with resource management, even though they sometimes derive divergent solutions from the different time intervals they consider. Second, well-intended environmental policies, like well-intended technologies, can have unintended adverse consequences. An example is the Delaney Clause, which was intended to prevent registration of carcinogenic pesticides and in the process favored more toxic compounds because only the least toxic materials could be applied at the sometimes unrealistically high doses needed to induce carcinogenesis. A third disadvantage is that the approach distracts attention from more meaningful environmental risks, including ones both related and unrelated to GMOs. The attention given to some putative direct effects, both by critics and by corporations that must respond, diverts resources from more complex processes that should receive more attention (Figure 13-1). The cycle of media attention resulting from such alarms, followed by subsequent dismissal of threats that turn out to be relatively unimportant, adds to the unwarranted impression that concerns over transgenics are collectively ill-founded.

At the other end of the spectrum is the view that deployment of GMOs should only be delayed where adverse consequences have been demonstrated. This approach has the advantage of insisting on scientifically based criteria. However, it has some significant disadvantages. First, it seriously underestimates the complexity of scaling across levels of biological organization, from small areas to landscapes, and from time scales based on rapid bioassays within grant cycles to time scales based on ecological and evolutionary processes. There is nothing scientific about ignoring the way that system processes change as you scale up, of which there are many examples in soil erosion, hydrology, engineering, economics, and biology. Second, the argument that risks should only be considered scientifically based once they have been "demonstrated" removes the proactive element of risk assessment and shifts the emphasis to remediation, which is far more difficult. This reactive approach has cost agrochemical companies dearly, with the result that they now invest substantial resources to prevent pesticide resistance. Finally, the argument that all risks can be classified as either "demonstrated" or "conjectural" is a false dichotomy that ignores the broad range of anticipated problems that we might term "realistic." I believe we can define "realistic risk" based on two criteria (Raffa et al. 1997): (a) the existence of relevant precedents that suggest likely outcomes, and (b) a proposed mechanism that can be delineated based on biologically verified processes. I will focus on two anticipated responses to pest resistance transgenes in tree plantations: evolution of resistant biotypes (Gould 1988; Raffa 1989) and alteration of complex ecosystem processes (Raffa et al. 1997; Obrycki et al. 2001).

## Biotype Evolution: Risks and Management

Biotype evolution is the emergence of new pest races that are no longer susceptible to a previously effective control tactic (Roush 1987,1997; Tabashnik 1994). The underlying mechanisms are well understood at several levels of biological organization. At the population level, biotype evolution is simply natural selection proceeding in an accelerated and directed fashion. A few individuals possessing fortuitous mutations survive, reproduce, and ultimately occur at disproportionately high frequencies. The physiological and biochemical mechanisms of insecticide resistance are also well understood

(Roush 1987, 1997; Tabashnik 1994). They include a variety of specific modalities involving altered behavior, detoxification, excretion, impermeability, and target site insensitivity.

Biotype evolution is not a rare phenomenon but is rather an inevitable consequence of any selection pressure that is applied continuously, uniformly, and at a sufficient intensity to cause high mortality (Gould 1988; Raffa 1989; Tiedje et al. 1989; Seidler and Levin 1994; Timmons et al. 1995; van Embden 1999). Resistant biotypes have evolved against all categories of pesticides, as well as resistant cultivars, biological control agents, sterile male releases, and cultural manipulations (Krysan et al. 1986; Raffa 1989). Resistant biotypes are prevalent among all taxa of insects, pathogens, and weeds. An example of the nearly infinite diversity of forms that biotype evolution can assume is provided from DDT sprays against malarial (*Anopheles*) mosquitoes. Euglossine bees that had been inadvertently exposed not only became DDT resistant but also incorporated this potent neurotoxin into their sexual behavior. Males now actively collect the granules after they are applied to houses and use them as a mating pheromone that enhances their attractiveness to females (Roberts et al. 1982). It is a remarkable example of how complex physiological and behavioral changes, including gender-specific behaviors, can evolve in a nontarget insect. In hindsight, the relevant pre-adaptation seems apparent: Euglossine males commonly gather pollen within flowers containing narcogenic phytochemicals, pass out in a stupor, awaken to exploit the elevated stud appeal that results from female attraction to these odors, and then repeat the process (Sazima et al. 1993).

The consequences of biotype evolution range from loss of efficacy to additional secondary effects. In most cases, loss of efficacy is the major and perhaps only significant consequence. Lost efficacy poses a substantial loss to the producer. Discovery, development, registration, and marketing of new pest control chemicals and transgenes are extremely expensive and marginally profitable processes. Corporations now invest heavily in biotype prevention tactics throughout all stages of agrochemical development to avoid these losses.

Effects that extend beyond reduced efficacy are of more concern from an environmental perspective. They include impacts on other resource managers, altered insect behavior, cross-resistance to other pesticides, and cross-resistance to natural plant defenses (Heinrichs and Mochida 1984; Fry 1989; Johnson and Gould 1992). The microbial insecticide *Bacillus thuringiensis* is currently the most widely used insect control agent in forestry. It is also used widely in tree fruit production. Its widespread acceptance arises from a combination of efficacy and economic and environmental attributes. *Bt* is sprayed on an as-needed basis only, which greatly diminishes the likelihood of biotype evolution. In contrast, uniform and continuous expression in trees could more rapidly select for resistant insects, which in turn could reduce the efficacy of an environmentally compatible tool that is also used in other cropping systems. There is substantial overlap in the species of insects that feed in forest, plantation, and tree-fruit systems.

An additional factor that needs to be considered is the possibility that resistant biotypes will display more damaging behaviors than the original genotypes. Altered behaviors sometimes arise in agroecosystems to which pesticides have been applied. For example, armyworms (*Spodoptera*) are important defoliators of many vegetable crops. Following repeated exposure to insecticides, some populations evolved a behavioral means of avoiding the toxins—increased boring into pepper fruits rather than feeding on leaves. Unfortunately these newly evolved populations inflicted greater economic losses than their untreated predecessors. Another significant threat of biotype evolution is cross-resistance. Resistance to one chemical class often confers resistance to widely unrelated compounds. An example occurred with the introduction of the synthetic pyrethroids. Although these materials were highly effective against laboratory and most field populations, they were almost immediately ineffective in certain regions. The pattern soon became obvious: Areas in which high levels of the organochlorine DDT had been applied harbored biotypes resistant to pyrethroids. DDT and pyrethroids are very dissimilar structurally but similar in their modes of action. Hence DDT-induced evolution of altered insect sodium channels conferred resistance against pyrethrum. Pyrethrum is a naturally occurring terpenoid present in *Chrysanthemum,* so from the insect's standpoint DDT is just a recent human variation on an ancient angiosperm theme. This raises the possibility that genetically engineered products might select for cross-resistance not only to other pesticides, but also to naturally occurring plant defenses. This possibility is not based on mere conjecture; increased ability to breach plant defense has already been observed in response to synthetic pesticides (Fry 1989).

The conditions that foster evolution of resistance biotypes are relatively well understood. In particular, the rate of evolution of pesticide resistance depends more on the pattern of application than on the mode of action (Roush 1987, 1997; Gould 1988; Tabashnik 1994). Patterns of expression that are uniform, large scale, unidirectional, and continuous select more rapidly for resistance than do intermittent patterns. If one achievement of pest management over the last 30 years ranks above all others, it is the replacement of calendar applications with targeted applications triggered by specific pest densities that have surpassed carefully defined economic thresholds. This change has brought an overall reduction in pesticide use, increased profits to the grower, increased durability of pesticides, improved pest control, and greater compatibility with diverse management tactics such as biological control, mating disruption, and cultural manipulations. Thus a particular technology, such as constitutive transgene expression in plants, may simultaneously represent a modern innovation at one level of scale and an archaic throwback at another. Conversely, many of the problems that ecologists identify can best be solved by molecular biologists or by molecular biologist–ecologist teams.

A number of tactics are available to manage biotype evolution, and many of them can be readily transferred to genetically modified plants. The tactics arise both from agriculture and from our understanding of naturally coevolved, stable plant-insect interactions. First, we should identify those systems in which biotype evolution appears particularly likely and avoid them. Expression of transgenic traits is more problematic with trees than with annual crops because of their long rotation times relative to insects. Many tree production systems provide crop-to-pest generation ratios an order of magnitude above that required to elicit pesticide resistance (Georghiou and Saito 1983). Thus, rapid rotation systems such as *Populus* and *Eucalyptus* are at less risk than longer rotation systems such as Douglas fir.

Second, biotypes are less likely to evolve when expression is variable over time. This is a key lesson obtained from the transition from calendar- to density-triggered pesticide applications. One opportunity to simulate threshold-triggered application is through wound-inducible expression, which is one of the most important means by which long-lived trees maintain stable defenses against herbivores and pathogens (Agrawal et al. 1999). Other approaches include limiting expression to certain periods of the growing season and to certain age categories. Various insect species show clear patterns of seasonal abundance, as well as association with particular age categories. Trees differentially allocate defenses according to these abundance patterns (Ikeda et al. 1977; Bingaman and Hart 1993).

Third, expression is most stable when it is limited in space. Spatial expression can vary among plant tissues, as is common in tree resistance mechanisms. There is also high variation among trees. Tactics that take advantage of this fact have already been employed with transgenic cotton and corn, in which fixed percentages of the "transgenic" seed are in fact not transgenes. Superimposing genetic diversity onto transgene expression adds another layer of protection. Hybrid poplars seem particularly amenable to this approach. For example, there is enormous genetic variation in poplar with regard to constitutive and induced resistance against a variety of pests; tolerance of defoliation, drought, and weeds; and intrinsic growth rates (Riemenschneider and McMahon 1993; Robison and Raffa 1994, 1997; Dunlap and Stettler 1996; Havill and Raffa 1999). Genotypic variation, coupled with variable deployment patterns, is a valuable safeguard against new insect and disease races and species, whether resistance is of transgenic or traditional origin (Peacock and Herrick 2000).

Fourth, the transgene should be compatible with other control methods. This helps maintain a diverse and opposing array of selective pressures. Natural systems often present insects with conflicting alternatives, which can provide an informative model for resistance management strategies (Raffa

1989). For example, aphids escape predation by releasing alarm pheromones when predators are near. Some wild potatoes react to aphid feeding by producing aphid alarm pheromone, which causes the aphids to drop from the plant (Gibson and Pickett 1983). Insects that became immune to this defense (i.e., ignored alarm pheromones) would presumably become more susceptible to predators.

Fifth, application of potentiators that specifically interfere with resistance mechanisms can greatly prolong activity. For example, insecticide resistance is often due to elevated detoxification enzyme titers. Inhibitors of P450s, such as piperonyl butoxide, interfere with detoxification and so can render resistant individuals susceptible. Again turning to coevolved systems as a model, some plants produce both insecticides and synergists. Altered herbivore-symbiont relationships offer another potential target. For example, the linear aminopolyol zwittermicin A, an antibiotic from soil bacteria, synergizes activity of *Bacillus thuringiensis* against gypsy moth (Broderick et al. 2000).

A sixth strategy is continued monitoring of resistance. As incipient resistance is identified, its mechanism and mode of inheritance should be characterized so that countermeasures can be employed (NRC 2002).

Experience with pesticides also suggests what is unlikely to work. For example, it seems reasonable that interrupted use of a pesticide will allow resistant populations to return to their original gene frequencies. The underlying assumption is that resistance characters are inherently disadvantageous and will be selected against in the absence of the pesticide. Unfortunately, that has proved not always to be the case (Roush 1987, 1997; Gould 1988). Resistance often becomes fixed within the genome. Moreover, the rationale that multiple genes of *Bt* are available, and so we should be able to stay ahead of resistance by deploying new ones as resistances evolve, is not supported by experience with other pesticides. Insects have shown remarkable ability to develop cross-resistance against very distant chemical groups (Georghiou and Saito 1983; Mullercohn et al. 1996). Cross-resistances have occurred among many classes of pesticides and by several mechanisms. Thus, it is no surprise that few pest management specialists are enthusiastic about this approach. Given the evolutionary history of insects in contending with millions of phytochemical combinations over hundreds of millions of years, placing our confidence in a few varieties of *Bt* does not seem the best option.

## Alteration of Complex Ecosystem Processes

The question of whether transgenic plants could exert serious environmental effects hinges greatly on whether gene flow into native plants occurs. If gene expression is limited to planted trees and mechanisms for preventing introgression are employed, then adverse environmental effects are likely to be limited and subject to remediation. However, if gene flow is likely, risk assessment must be tailored accordingly.

The issue of potential gene escape has generated much debate among geneticists (e.g., Regal 1993; Linder and Schmitt 1995; Strauss et al. 1995; Timmons et al. 1995; Kareiva et al. 1996; Paoletti and Pimentel 1996; Parker and Kareiva 1996). However, there is general agreement that there is some gene flow from cultivated to native and feral plants. Some potential mechanisms include hybridization, dissemination of vegetative material, and vectors. Strauss et al. (1995) concluded that "gene flow within and among tree populations is usually extensive, which makes the probability of transgene escape from plantations high" (Adams 1992; Raybould and Gray 1993). They based their conclusions on "high rates of gene dispersal by pollen and seed, and proximity of engineered trees to natural or feral stands of interfertile species."

Proposed adverse environmental effects include enhanced weediness, reduced biodiversity, and alteration of ecosystem processes (Obrycki et al. 2001). Moreover, Parker and Kareiva (1996) identified pest resistance as the transgenic property most likely to cause such problems. Enhanced weediness has received much attention, and so I will not address it here. Likewise, the issue of biodiversity has been discussed in detail elsewhere; space considerations here do not allow me to add to that body of work. I will focus instead on ecosystem processes, particularly examples in which prior experience with pesticides and new cultivars have resulted in exacerbated pest problems. These include direct and indirect effects on natural enemies and release of competitors.

The extent to which transgenic trees will directly affect predators depends on the gene product. In particular, stable materials are more likely to be biomagnified across trophic levels. Thus, substantial biomagnification is not known to occur with *Bt*. For the future, however, three considerations deserve attention: First, herbivores usually evolve resistance to xenobiotics more rapidly than do predators. This appears to relate to their coevolutionary history with plant defense chemicals (Berenbaum et al. 1996). Second, even if a xenobiotic has equivalent toxicity to an herbivore and its predator, it will have a greater effect on predator than herbivore populations. This has been demonstrated in numerous mathematical models and relates to issues of prey finding, prey handling, prey survival, and differences in reproductive capacity. Third, control methods that reduce generalist predators or parasitoids of target pests can cause population increases of non-pest species. Because parasites typically develop better in healthy than in weakened hosts (Visser 1994; Havill and Raffa 2000), anything that reduces herbivore vigor is likely to cause premature parasite death. Hence, transgenic plantings can become lethal sinks for parasite populations (Johnson and Gould 1992). Such dynamics of secondary pest flare-ups are commonly seen with sprayed toxins.

Emergence of secondary pests is a common and serious problem when insecticides or resistant cultivars are deployed. For example, mite outbreaks often follow applications of pyrethroids and historically followed application of DDT. The most serious outbreaks of spruce spider mite followed aerial application of DDT against western spruce budworm (Furniss and Carolin 1980). Likewise, introductions of new cultivars have resulted in the emergence of previously unimportant pests (Harlan 1980; Oka and Bahagiawati 1984). A new cultivar bred for pest resistance sometimes reduces that pest species but frees others from competition. In other cases, breeding for unrelated properties has inadvertently enhanced the plant's susceptibility or nutritional suitability to another insect. Again, natural systems demonstrate that these are not rare examples. For almost every phytochemical that has been shown to confer resistance against insects, there are other insect species that use the same compound for nutrition, defense, or communication (Rosenthal and Berenbaum 1992). The seriousness of this problem again depends on whether there is gene flow from transgenic plants into wild and feral populations.

## Risk Assessment: Mechanisms and Precedents

Several models have been proposed for assessing risks from transgenic pest-resistant plants. Each has certain advantages and disadvantages, and their relative applicability, in my view, depends on system properties and sterility mechanisms.

### Pesticides

Using pesticide registration as a model for pest-resistant transgenes has several advantages. First it focuses attention on the gene product, which can be tested like other compounds for acute and chronic toxicity to humans, for effects on beneficial organisms, and alone or in combination with phytochemicals. Such tests are expensive, but the methods are straightforward, experimental conditions can be shielded from investigator bias, and the results can be interpreted relatively easily. I do not minimize the difficulty of conducting toxicological studies, and I recognize that there are sometimes peculiar dose–response relationships, complex immunological interactions, and unpredictable multichemical interactions. But corporations welcome this approach because it gives them a clear target, and they can draw on years of experience with agrochemicals, pharmaceuticals, industrial reagents, and cosmetics. Researchers in all sectors want to be sure that materials they introduce into the environment are safe. Whether this approach is adequate depends largely on whether transgenes have the potential to become established in wild populations. If sterility is incorporated into the genome and vegetative parts are contained, there are no apparent risks beyond those associated with any other environmental input. If those conditions are not met, however, then the pesticide analogy is applicable to estimating direct effects but inadequate for evaluating environmental risks arising from more complex processes. Pesticide inputs can be halted at any time, and materials with excessively long residual periods are banned. This is a valuable aspect of pesticides, as compounds that had been approved for many decades commonly lose their certification as new toxicological problems are discovered.

## Pest-Resistant Cultivars

Transgenic plants are often compared with plants bred by traditional methods. This analogy has some value in that it recognizes that environmental risks apply to all monocultures, regardless of how the genome was derived. The comparison is sometimes extended to emphasize the added advantage of genes' being introduced in a targeted rather than random fashion. From an efficacy standpoint that is indeed a major benefit to genetic engineering. But when it is extrapolated to environmental risk, it reverses the product-not-process principle—suddenly the process becomes the proposed basis for justification. At the scale of pest management, the only thing transgenic and traditionally bred crops have in common is the process, that is, deploying the toxin through plants. As such, equating transgenes with traditionally bred resistant plants is a significant deviation from the product-not-process principle. It is also sometimes argued that a single gene would have less effect than the multiple rearrangements that take place through traditional breeding. However there is little evidence to support that assumption, and there are sound biological bases for being skeptical (NRC 2002). First, single genes often have dramatic effects, as is the case with many human diseases, pesticide-resistant insects and pathogens, and cultivar-resistant pathogens. Secondly, the genes that can be deployed through traditional breeding are limited to the pool available from the plant's evolutionary history, as opposed to the taxonomic novelty of many transgenes. This constitutes a strong argument for the efficacy and utility of genetic engineering but not for its environmental safety. Our most severe pest problems have occurred in non-coadapted systems. Consequences of taxonomically unrelated genes, including single genes, may vary significantly, depending on the genomic, physical, and biological environment into which they are introduced. We currently lack the ability to predict such general trends (NRC 2002). Third, the analogy does not consider either pleiotropic effects or gene by environment interactions. Yet we know these can be extremely important in determining how insect–plant interactions and insect populations behave. In conclusion, resistance breeding is a useful but inadequate precedent.

### Planned Introductions of Beneficial Organisms

A third regulatory area with applicability to transgenic pest resistance is the planned introduction of beneficial organisms. This approach has several merits in systems where the possibility of gene escape cannot be excluded. First, it is consistent with the product-not-process philosophy. The 1987 National Academy of Sciences report stated that introducing GMOs "poses no risks *different from* the introduction of unmodified organisms" (emphasis mine). Note that it did not say *"poses risks less than."* To argue that GMOs should be treated differently from other introduced organisms would be a radical departure from the product-not-process guideline. Second, risk assessment of biological control agents places a premium on ecological considerations, specifically on how the agents will affect other components of the ecosystem. This is precisely the area where risk assessment of GMOs has lagged. Third, treating GMOs as other putative biocontrol agents does not pose insurmountable obstacles to registration and commercialization. Literally hundreds of biocontrol agents have been approved for release. Moreover it is disingenuous for proponents of GMOs to argue that they have been singled out for unprecedented scrutiny, while at the same time objecting to risk assessment standards that have been applied to other sectors for many decades. Fourth, the biological control framework seems more workable than the framework of invasive species, which has likewise been proposed as a model for GMOs. GMOs and biological control agents are both selected because of specific, desirable properties. In contrast, most of our most damaging nonindigenous species were not introduced deliberately but rather have properties that contribute to enhanced "invasiveness."

The biological control analogy thus also has limitations. It might be argued that the standards applied to biological control agents are too lax, as some adverse effects have occurred. However most such problems have arisen from organisms that were released privately (i.e., without regulatory oversight) or were released many years ago under standards less well informed than those applied at present. Another limitation is that biological control agents are intended to become self-sustaining, whereas in the case of transgenic trees, that would be an unintended consequence.

## Conclusions

Transgenic expression provides some unique opportunities for pest resistance in plantation trees that cannot be achieved easily by other means. These opportunities include reduction of pesticide inputs, increased productivity and profitability, increased feasibility of biofuels, and reduced pressures on multiple-use forests. There are also some risks, and evaluating them with trees is especially difficult and still in its infancy. The difficulties relate to the long time scale of host, relative to pest, generation times; the large spatial scale of some plantation systems; and the multiple uses and ecological functions of adjoining conspecific plantation and native trees. The consideration of economic and environmental opportunities potentially lost makes risk assessment all the more difficult. Experiences in resistance management of pesticides, as well as risk assessment with the planned introduction of biological control agents, provide useful approaches.

Some adverse consequences could exert environmental harm, and others would squander valuable transgenes. The reasonable likelihood of risks can be identified where two conditions are met: relevant precedent and validated mechanisms. Indirect ecological processes are the most likely to yield adverse effects, yet these are the least likely to be identified by current programs.

Several obstacles impede scientifically based risk assessment, including (a) inconsistent application of the product-not-process principle, by both opponents and proponents of GMOs; (b) extrapolation across multiple levels of biological organization, spatial scales, and time frames without verification of the underlying assumption that system processes remain uniform; (c) insufficient attention to general principles in cases when specific information is not yet available; and (d) systematic biases in research support that favor a risk analysis process based on precise but simplified conditions over complex but more realistic interactions.

The extent to which various risk assessment approaches apply depends on the extent to which gene flow can be eliminated. Where sterility is introduced and vegetative materials are contained, the procedures used for pesticide evaluation are most applicable. Where either of those ingredients is missing, the most applicable standard is the planned introduction of putatively beneficial organisms. A major challenge lies in the inability of ecologists to quantify the likelihood of various potential adverse effects, given the requisite long time frames, complex trophic interactions, and extensive spatial scales. Thus, there is a significant need to develop effective models that could help bridge the gaps between transgene development and realistic environmental assessment.

# Acknowledgment

This manuscript is derived from an oral presentation to the IUFRO Conference on Somatic Cell Genetics and Molecular Genetics of Trees, in July 2001. I greatly appreciate the invitation of Steve Strauss and Toby Bradshaw to present my thoughts, and to all of the attendees for their valuable insights. Support was provided by IUFRO and the University of Wisconsin-Madison College of Agricultural and Life Sciences. The helpful reviews of Jaimie Powell (Department of Entomology, University of Wisconsin-Madison), Steve Strauss, and three anonymous reviewers are greatly appreciated.

# References

Adams, W.T. 1992. Gene Dispersal within Forest Tree Populations. *New Forestry.* 6: 217–240.

Agrawal, A., S. Tuzon, and E. Bent. 1999. *Induced Plant Defenses against Herbivores and Pathogens.* St. Paul, MN: APS Press.

Barrett, G.W., and A. Farina. 2000. Integrating Ecology and Economics. *BioScience* 50: 311–312.

Baskin, Y. 1999. Yellowstone Fires: A Decade Later—Ecological Lessons Learned in the Wake of the Conflagration. *BioScience* 49: 93–97.

Berenbaum, M.R., C. Favret, and M.A. Schuler. 1996. On Defining Key Innovations in an Adaptive Radiation—Cytochrome p450s and Papilionidae. *American Naturalist* 148: S139–S155.

Bingaman, B.R., and E.R. Hart. 1993. Clonal and Leaf Age Variation in *Populus* Phenolic Glycosides: Implications for Host Selection by *Chrysomela scripta* (Coleoptera: Chrysomelidae). *Environmetal Entomology* 22: 397–403.

Broderick, N.A., R.M. Goodman, K.F. Raffa, and J. Handelsman. 2000. Synergy between *Zwittermicin A* and *Bacillus thuringiensis* subsp. *kurstaki* against Gypsy Moth (Lepidoptera: Lymantriidae). *Environmental Entomology* 29: 101–107.

Canadian Forest Service. 1994. *Forest Depletions Caused by Insects and Diseases in Canada 1982–1987.* Forest Insect and Disease Survey. Ottawa: Canadian Forest Service, Natural Resources Canada.

Crooks, K.R., and M.E. Soulé. 1999. Mesopredator Release and Avifaunal Extinctions in a Fragmented System. *Nature* 400: 563–566.

Dunlap, J.M., and R.F. Stettler. 1996. Genetic Variation and Productivity of *Populus trichocarpa* and Its Hybrids: Phenology and *Melampsora* Rust Incidence of Native Black Cottonwood Clones from Four River Valleys in Washington. *Forest Ecology and Management* 87: 233–256.

Estes, J.A. 1996. Predators and Ecosystem Management. *Wildlife Society Bulletin* 24: 390–396.

Estes, J.A., K. Crooks, and R. Holt. 2001. Predators, Ecological Role of. *Encyclopedia of Biodiversity* 4: 857–878.

Fry, J.D. 1989. Evolutionary Adaptation to Host Plants in a Laboratory Population of the Phytophagous Mite *Tetranychus urticae* Koch. *Oecologia* 81: 559–565.

Furniss, R.L., and V.M. Carolin. 1980. *Western Forest Insects.* Misc. Publ. 1339. Washington, DC: USDA Forest Service

Gabrielle, J.P., and J.N. Siedow. 1999. *Applications of Biotechnology to Crops: Benefits and Risks.* Council for Agricultural Science and Technology Issue paper 12.

Georghiou, G.P., and T. Saito. 1983. *Pest Resistance to Pesticides.* New York: Plenum Press.

Gibson, R.W., and J.A. Pickett. 1983. Wild Potato Repels Aphids by Release of Aphid Alarm Pheromone. *Nature* 302: 608–609.

Gould, F. 1988. Evolutionary Biology and Genetically Engineered Crops. *BioScience* 38: 26–33.

Harlan, J.R. 1980. Origins of Agriculture and Crop Evolution. In *Biology and Breeding for Resistance to Arthropods and Pathogens in Agricultural Plants,* edited by M.K. Harris. College Station: Texas A & M University Press, 1–8.

Havill, N.P., and K.F. Raffa. 1999. Effects of Eliciting Treatment and Genotypic Variation on Induced Resistance in *Populus*: Impacts on Gypsy Moth Development and Feeding Behavior. *Oecologia* 120: 295–303.

Havill, N.P., and K.F. Raffa. 2000. Compound Effects of Induced Plant Responses on Insect Herbivores and Parasitoids: Implications for Tritrophic Interactions. *Ecological Entomology* 25: 171–179.

Heinrichs, E.A., and O. Mochida. 1984. From Secondary to Major Pest Status: The Case of Insecticide-Induced Rice Brown Planthopper, *Nilaparvata lugens*, Resurgence. *Protection Ecology* 7: 201–218.

Henke, S.E., and F.C. Bryant. 1999. Effects of Coyote Removal on the Faunal Community in Western Texas. *Journal of Wildlife Management* 63: 1066–1081.

Ikeda, T., F. Matsumura, and D.M. Benjamin. 1977. Chemical Basis for Feeding Adaptation of Pine Sawflies *Neodiprion rugifrons* and *Neodiprion swainei*. *Science* 197: 497–499.

Johnson, M.T., and F. Gould. 1992. Interaction of Genetically Engineered Host Plant Resistance and Natural Enemies of *Heliothis virescens* (Lepidoptera, Noctuidae) in Tobacco. *Environmental Entomology* 21: 586–597.

Kareiva, P., I.M. Parker, and M. Pascual. 1996. Can We Use Experiments and Models in Predicting the Invasiveness of Genetically Engineered Organisms? *Ecology* 77: 1670–1675.

Kleiner, K.W., D.D. Ellis, B.H. McCown, and K.F. Raffa. 1995. Field Evaluation of Transgenic Poplar Expressing *Bacillus thuringiensis* d-endotoxin Gene against Forest Tent Caterpillar (Lepidoptera: Lasiocampidae) and Gypsy Moth (Lepidoptera: Lymantriidae). *Environmental Entomology* 24: 1358–1364.

Krysan, J.L., D.E. Foster, T.F. Branson, K.R. Ostlie, and W.S. Cranshaw. 1986. Two Years before the Hatch: Rootworms Adapt to Crop Rotation. *Bulletin of the Entomological Society of America* 32: 250–253.

Linder, C.R., and J. Schmitt. 1995. Potential Persistence of Escaped Transgenes: Performance in Transgenic, Oil-Modified *Brassica* Seeds and Seedlings. *Ecological Applications* 5: 1056–1068.

Longland, W.S. 1991. Risk of Predation and Food Consumption by Black-Tailed Jackrabbits. *Journal of Range Management* 44: 447–450.

Losey, J.E., L.S. Raynor, and M.E. Carter. 1999. Transgenic Pollen Harms Monarch Larvae. *Nature* 399: 214.

McCown, B.H., D.E. McCabe, D.R. Russell, D.J. Robison, K.A. Barton, and K.F. Raffa. 1991. Stable Transformation of *Populus* and Incorporation of Pest Resistance by Electrical Discharge Particle Acceleration. *Plant Cell Reports* 5: 590–594.

Meeusen, R.L., and G. Warren. 1989. Insect Control with Genetically Engineered Crops. *Annual Review of Entomology* 34: 373–381.

Mullercohn, J., J. Chaufaux, C. Buisson, N. Gilois, V. Sanchis, and D. Lereclus. 1996. *Spodoptera littoralis* (Lepidoptera, Noctuidae) Resistance to CRYic and Cross-Resistance to Other *Bacillus thuringiensis* Crystal Toxins. *Journal of Economic Entomology* 89: 791–797.

Naiman, R.J., G. Pinay, C.A. Johnston, and J. Pastor. 1994. Beaver Influences on the Long Term Biogeochemical Characteristics of Boreal Forest Drainage Networks. *Ecology* 75: 905–921.

NAS (National Academy of Sciences). 1987. *Committee on the Introduction of Genetically Engineered Organisms into the Environment.* Washington, DC: NAS Press.

Naveh, Z., and A.S. Lieberman. 1994. *Landscape Ecology: Theory and Application.* 2nd ed. New York: Springer-Verlag.

NRC (National Research Council). 2002. *Environmental Effects of Transgenic Plants: The Scope and Adequacy of Regulation.* Washington, DC: NRC Press.

Obrycki, J.J., J.E. Losey, O.R. Taylor, and L.C.H. Jesse. 2001. Transgenic Insecticidal Corn: Beyond Insecticidal Complexity to Ecological Complexity. *Bioscience* 51: 353–361.

Oka, F.N., and A.H. Bahagiawati. 1984. Development and Management of a New Brown Planthopper (*Nilaparvata leguns* Stal) Biotype in North Sumatra. *Indonesia Contributions from the Central Research Institute for Food Crops Bogor* 71: 1–33.

Paoletti, M.G., and D. Pimentel. 1996. Genetic Engineering in Agriculture and the Environment. *Bioscience* 46: 665–673.

Parker, I.M., and P Kareiva. 1996. Assessing the Risks of Invasion for Genetically Engineered Plants: Acceptable Evidence and Reasonable Doubt. *Biological Conservation* 78: 193–203.

Peacock, L., and S. Herrick. 2000. Responses of the Willow Beetle *Phratora vulgatissima* to Genetically and Spatially Diverse *Salix* spp. Plantations. *Journal of Applied Ecology* 37: 821–831.

Pollach, M.M., R.J. Naiman, H.E. Erickson, C.A. Johnson, J. Pastor, and G. Pinaj. 1995. Beavers vs. Engineers: Influences on Biotic and Abiotic Characteristics of Drainage Basins. In *Linking Species and Ecosystems,* edited by C.G. Jones and J.H. Lawton. New York: Chapman and Hall, 117–126.

Raffa, K.F. 1989. Genetic Engineering of Trees to Enhance Resistance to Insects: Evaluating the Risks of Biotype Evolution and Secondary Pest Outbreak. *BioScience* 39: 524–534.

Raffa, K.F., K.W. Kleiner, D.D. Ellis, and B.H. McCown. 1997. Environmental Risk Assessment and Deployment Strategies in Genetically Engineered Insect-Resistant *Populus.* In *Micropropagation, Genetic Engineering, and Molecular Biology of Populus,* edited by N.B. Klopfenstein, Y.W. Chun, M.-S. Kim, and M.R. Ahiya. General Technical Report, Rocky Mountain Forest and Range Experimental Station. Fort Collins, CO: USDA Forest Service, 249.

Raybould, A.F., and A.J. Gray. 1993. Genetically Modified Crops and Hybridization with Wild Relatives: A UK Perspective. *Journal of Applied Ecology* 30: 199–219.

Regal, P.J. 1993. The True Meaning of "Exotic Species" as a Model of Genetically Engineered Organisms. *Experientia* 49: 225–234.

Riemenschneider, D.E., and B.G. McMahon. 1993. Genetic Variation among Lake States Balsam Poplar Populations Is Associated with Geographic Origin. *Forest Science* 39: 130–136.

Roberts, D.R., W.D. Alecrim, J.M. Heller, S.R. Ehrhardt, and J.B. Lima. 1982. Male *Eufriesia purpurata,* a DDT-Collecting Bee in Brazil. *Nature* 297: 62–63.

Robison, D.J., and K.F. Raffa. 1994. Characterization of Hybrid Poplar Clones for Resistance to the Forest Tent Caterpillar. *Forest Science* 40: 686–714.

———. 1997. Effects of Constitutive and Inducible Traits of Hybrid Poplars on Forest Tent Caterpillar Feeding and Population Ecology. *Forest Science* 43: 252–267.

Romme, W.H., and D. Knight. 1981. Fire Frequency and Subalpine Forest Succession along a Topographic Gradient in Wyoming. *Ecology* 62: 319–326.

Rosenthal, G.A., and M.R. Berenbaum. 1992. Herbivores: Their Interactions with Secondary Plant Compounds. New York: Academic Press.

Roush, R.T. 1987. Ecological Genetics of Insecticide and Acaricide Resistance. *Annual Review of Entomology* 32: 361–380.

———. 1997. Managing Resistance to Transgenic Crops. Chap. 15 in *Advances in Insect Control: The Role of Transgenic Plants,* edited by Nadine Carozzi and Michael Koziel. Bristol, PA: Taylor and Francis, 271–294.

Sazima, M., S. Vogel, A. Cocucci, and G. Hausner. 1993. The Perfume Flowers of *Cyphomandra* (Solanaceae): Pollination by Euglossine Bees, Bellows Mechanism, Osmophores, and Volatiles. *Plant Systematics and Evolution* 187: 51–88.

Seidler, R.J., and M. Levin. 1994. Potential Ecological and Nontarget Effects of Transgenic Plant Gene Products on Agriculture, Silviculture, and Natural Ecosystems: General Introduction. *Molecular Ecology* 3: 1–3.

Strauss, S.H., S.A. Knowe, and J. Jenkins. 1997. Benefits and Risks of Transgenic Trees. *Journal of Forestry* 95: 12–19.

Strauss, S.H., K.F. Raffa, and P.C. List. 2000. Ethical Guidelines for Using Genetically Engineered Trees in Forestry. *Journal of Forestry* 98: 47–48.

Strauss S.H., W.H. Rottmann, A.M. Brunner, and L.A. Sheppard. 1995. Genetic Engineering of Reproductive Sterility in Trees. *Molecular Breeding* 1: 5–26.

Tabashnik, B.E. 1994. Evolution of Resistance to *Bacillus thuringiensis. Annual Review of. Entomology* 39: 47–79.

Tiedje, J.M., R.K. Colwell, Y.L. Crossman, R.E. Hodson, R.E. Lenski, R.N. Mack, and P.J. Regal. 1989. The Planned Introduction of Genetically Engineered Organisms: Ecological Considerations and Recommendations. *Ecology* 70: 298–315.

Timmons, A.M., Y.M. Charters, S.J. Dubbels, and M.J. Wilkinson. 1995. Assessing the Risks of Wind Pollination from Fields of Genetically Modified *Brassica napus* spp. *oleifera. Euphytica* 85: 417–423.

Turner, M.G., R.H. Gardner, and R.V. O'Neill. 2001. *Landscape Ecology in Theory and Practice: Pattern and Process.* New York: Springer.

Turner, M.G., W.H. Romme, R.H. Gardner, and W.W. Hargrove. 1997. Effects of Fire Size and Pattern on Early Succession in Yellowstone National Park. *Ecological Monographs* 67: 411–433.

van Emden, H.F. 1999. Transgenic Host Plant Resistance to Insects—Some Reservations. *Annals of the Entomological Society of America* 92: 788–797.

Visser, M.E. 1994. The Importance of Being Large—The Relationship between Size and Fitness in Females of the Parasitoid *Aphaereta minuta* (Hymenoptera, Braconidae). *Journal of Animal Ecology* 63: 963–978.

Wagner, F.H., and L.C. Stoddard. 1972. Influence of Coyote Predation on Black-Tailed Jackrabbit Populations in Utah. *Journal of Wildlife Management* 36: 329–342.

# Index